Warwickshire School of Nu
Nurse Education Centre
Central Hospital
Near Warwick, Warwickshi

G000139649

**This book is to be returned on or before
the last date stamped below.**

Book No.   10M   2580   S

Warwickshire Area Health Authority
WARWICKSHIRE SCHOOL OF NURSING

Books are on loan for 28 days only. They may
be renewed for one further period of 28 days.

The Drugs Handbook

Paul Turner
Professor of Clinical Pharmacology, St. Bartholomew's Hospital, London EC1

and

Glyn N. Volans
Consultant Clinical Pharmacologist and Deputy Director of the Poisons Unit, Guy's Hospital, London SE1

© Paul Turner and Glyn N. Volans 1978

Reprinted 1978

All rights reserved. No part of this
publication may be reproduced or
transmitted, in any form or by any means,
without permission

Published by
THE MACMILLAN PRESS LTD
London and Basingstoke
Associated companies in New York
Dublin Melbourne Johannesburg and Madras

Printed in Great Britain by
John Wright & Sons Ltd., at
The Stonebridge Press, Bristol

Sold subject to the standard conditions of
the Net Book Agreement

British Library Cataloguing in Publication Data
Turner, Paul, b. 1933
    The Drugs Handbook.
    1. Drugs—Catalogs
    I. Title II. Volans, Glyn N
    615'. 1'0941   RS355

    ISBN 0–333–21612–1
    ISBN 0–333–23655–6 Pbk

INTRODUCTION

Patient care has now extended far beyond the patient–doctor relationship and involves several different highly trained health care professions, including amongst others nurses, midwives, pharmacists, occupational therapists, physiotherapists, radiographers, electro-encephalographers, dieticians, social workers, psychologists and medical secretaries. Although prescribing of medicines is the responsibility of a doctor, the drugs which medicines contain may influence patients in many ways, and it is important that others involved in a patient's health care should have ready access to information on the various medicines which he or she may receive, either by prescription or by over-the-counter purchase. The primary purpose of this book is to provide such information for these and other groups within the health professions. It is not a textbook of clinical pharmacology or medical treatment nor is it intended to be: rather it is meant to be a short guide to the mechanism of action, therapeutic indications and chief unwanted effects of most medicines available in the United Kingdom.

Drugs and Medicines
A doctor usually prescribes a *drug*, but the patient receives a *medicine*. The medicine is the whole formulation in which the drug, the active substance, is combined with other ingredients to form a convenient form of administration such as a tablet, capsule, suppository, inhalation, ointment or injection. We have not included all constituents of the medicines listed in this book, but have mentioned only those substances which we believe may contribute to the therapeutic or adverse effects of the medicine involved. It must be stressed that the mention of a medicine in this book, and statements about its indications, do not imply that it is necessarily an effective treatment, or that the authors believe it to be such; in fact we believe that for large numbers of drugs there is no good evidence of their effectiveness.

Names of Drugs
Most drugs have at least three names. The first is the full chemical name, which is too long and complicated to use regularly. More convenient is the shorter *generic* or *approved* name, which may become accepted internationally. Finally, there is the *brand* or *trade* name, given by the pharmaceutical manufacturer for its own particular brand or formulation. For example, 4-amino-5- chlor-N-(2-diethylaminoethyl)-2-methoxybenzamide hydrochloride is the chemical name for metaclopramide hydrochloride, the approved name of the drug marketed at present by at least two drug firms under the brand names Maxolon and Primperan. We have distinguished between approved and brand names by compiling two separate alphabetical lists. The main body of the book is devoted to approved names, with a brand name index at

the end, and to save space and avoid duplication of information almost all the brand names have been cross-referenced to the appropriate approved names. Also in the interests of saving space, where a number of drugs have essentially similar effects we have cross-referenced approved names to those which we consider to be the most typical and most frequently prescribed. Thus all cross-references refer to the list of approved names, and are shown in bold type, e.g. **amethocaine.**

Drugs in Pregnancy
Several drugs are known to be hazardous to the developing foetus and for most drugs there is no definite information on their safety in pregnancy. We have avoided constant repetition of this but would stress that in pregnancy all drugs should be used with caution and only when essential.

Technical Terms, Further Information
This book has not been written for the general public, and it assumes a basic knowledge and understanding of human biology and disease. Many readers may, nevertheless, wish to refer to larger books for more complete and detailed information, and we would recommend:
1. For information on drugs:
   Turner, P. and Richens, A. *Clinical Pharmacology.* 2nd Edition, 1975. Churchill–Livingstone, London.
   Laurence, D. R. *Clinical Pharmacology.* 4th Edition, 1973. Churchill–Livingstone, London.
   Goodman, L. S., and Gilman, A. (Eds.) *The Pharmacological Basis of Therapeutics.* 1975. Ballière Tindall, London.
   Blacow, N. W. (Ed.) Martindale: *The Extra Pharmacopoeia.* 27th Edition, 1977. Pharmaceutical Press, London.
2. For information on diseases and their management:
   Houston, J. C., Joiner, C. L. and Trounce, J. R. *A Short Textbook of Medicine.* 5th Edition, revised 1977. Hodder and Stoughton, London.
   Mann, W. N. (Ed.) Conybeare's *Textbook of Medicine.* 16th Edition, 1975. Churchill–Livingstone, London.
3. For information on treatment of poisoning:
   Matthew, H. and Lawson, A. H. *Treatment of Common Acute Poisonings.* 3rd Edition, 1975. Churchill–Livingstone, London.

July 1977

PAUL TURNER
GLYN N. VOLANS

ACKNOWLEDGEMENTS
The authors would like to express their thanks and gratitude to Miss B. Conaty for her help in compiling the lists of drugs and for typing the manuscript.

Three abreviations have been used throughout the book:
(b) indicates a borderline substance, i.e. a substance which may only be prescribed as a drug under certain conditions.
(c) indicates a drug whose prescription is controlled under the Misuse of Drugs Act.
(d) discontinued by the manufacturers during the year prior to publication. As supplies will still be available from pharmacies until stocks run out, these products have not yet been deleted from our lists.

GROUP NAMES: DEFINITIONS

aminoglycoside   A drug with a chemical structure related to **streptomycin.**
anabolic   Stimulates cell metabolism causing increased tissue growth.
analgesic   Relieves pain.
antacid   Neutralises acid produced by the stomach.
antagonist   Opposing the action or blocking the effect of another drug.
anticholinergic   Blocks the action of **acetylcholine** or cholinergic (acetylcholine-like) drugs.
anticholinesterase   Naturally occurring enzyme which breaks down **acetylcholine** and thus brings its action to an end.
anticoagulant   Precents blood from clotting.
anticonvulsant   Stops or prevents epileptic seizures.
antiemetic   Prevents nausea and vomiting.
antihypertensive   Lowers blood pressure.
antipyretic   Lowers body temperature in febrile conditions.
antispasmodic   Relieves spasm of muscles, e.g. in the gastro-intestinal tract (gastro-intestinal colic).
anxiolytic   Reduces anxiety.
astringent   Precipitates protein to form a protective layer over damaged skin or mucous membranes.
bactericidal   Kills bacteria.
bacteriostatic   Inhibits growth of bacteria but does not kill them.
bronchodilator   Increases the diameter of the airways in the respiratory system and thus reduces the physical resistance to breathing.
cardioselective   Having actions on the heart without the other effects usually found in drugs of a particular group, e.g. beta-adrenoceptor blocking drugs. In practice the cardioselectivity is usually relative and the other effects can be traced, although in a less pronounced form.

carminative   A drug which facilitates the eructation of gas from the stomach.

cholinergic   Has actions similar to **acetycholine.**

corticosteroids   Hormones (natural or synthetic) with actions on metabolism and against tissue inflammation. The natural hormones are produced by the adrenal gland.

cytotoxic   Has toxic effects upon living cells which reduce growth or cause destruction of the cells.

diuretic   Increases urine output.

emetic   Induces vomiting.

expectorant   Aids removal of sputum from the lungs and respiratory passages.

fibrinolytic   Dissolves or otherwise destroys the fibrin which is formed when blood clots.

G6PD   Glucose 6-phosphate dehydrogenase: an enzyme involved in carbohydrate metabolism. Some patients exhibit an inherited deficiency of this enzyme and are thus more susceptible to certain diseases and adverse drug effects.

haemostatic   Stops bleeding and prevents blood loss.

hormone   A naturally occurring substance which is secreted by a gland into the blood stream, whence it is carried to the part of the body on which it acts. Insulin, for example, is secreted by the pancreas and acts at sites all over the body.

hypnotic   Facilitates sleep.

mucolytic   Alters mucus within the respiratory system.

narcotic analgesic   Pain-relieving drug of the opium group. Liable to have addictive properties.

neurotransmitter   A biochemical substance which acts in the transmission of nerve impulses.

ototoxic   Causing damage to nerves involved in hearing.

phenothiazine   A drug with a chemical structure similar to **chlorpromazine.**

purgative   Facilitates evacuation of the bowels.

rubefacient   Causes reddening of the skin.

sedative   Reduces arousal.

sympathomimetic   Has actions similar to the sympathetic chemical transmitter of nervous system (adrenaline and noradrenaline).

thiazide   A drug with a chemical structure similar to **chlorothiazide.**

tranquilliser   Drugs with sedative actions on the brain which are used in the treatment and management of certain psychiatric disorders, e.g. schizophrenia, mania.

vasoconstrictor   Constricts blood vessels.

vasodilator   Dilates blood vessels.

Part I

Approved Names

A

Acebutalol
Beta-adrenoceptor blocking drug, with limited cardioselectivity. Uses, side effects, etc. as for **Propranolol.**

Acepifylline
Bronchodilator with actions similar to **Aminophylline.**

Acetaminophen
U.S.A.: see **Paracetamol.**

Acetarsol
Arsenical agent used in treatment of gut amoebiasis or vaginal trichomonas. Given by mouth, pessary or suppository. Adverse effects include vomiting, diarrhoea, rashes. **Dimercaprol** is antidote.

Acetazolamide
Weak diuretic. Also used in glaucoma to reduce intra-ocular pressure, and as an anticonvulsant. Acts by inhibiting carbonic anhydrase and so reduces hydrogen ions available for exchange with sodium ions. May cause drowsiness, mental confusion and paraesthesiae.

Acetohexamide
Oral antidiabetic agent with actions and uses similar to **Chlorpropamide.**

Acetomenaphthone
See **Vitamin K.**

Acetophenetidin
U.S.A.: see **Phenacetin.**

Acetylcholine
Neurotransmitter, particularly in parasympathetic system. Peripheral effects include miosis, paralysis of accommodation, increased glandular secretions, contraction of smooth muscle in gastro-intestinal, respiratory and urogenital systems, slowing of heart, vasodilatation. These effects blocked by **Atropine sulphate.** Not used clinically.

Acetylcysteine
Mucolytic. Administered by inhalation from a nebuliser. Liquefies mucus and aids expectoration in diseases where mucus is troublesome, e.g. chronic bronchitis. May cause bronchospasm, haemoptysis, nausea and vomiting.

3

Ace

Acetylsalicylic acid (Aspirin)
Anti-inflammatory, antipyretic analgesic. Inhibits prostaglandin synthesis. May cause gastric erosion and haemorrhage, hypersensitivity reactions. Tinnitus, hyperventilation leading to respiratory and cardiovascular failure in overdose. Interacts with oral anticoagulants and sulphonylureas.

Actinomycin D
Cytotoxic antibiotic used in neoplastic disease. Adverse effects include bone marrow depression.

Adenosine Monophosphoric Acid (AMP)
Source of high-energy phosphate bonds for tissue metabolism. Suggested for use in cardiovascular disease and rheumatism but efficacy unproven.

Adenosine triphosphoric acid (ATP)
Source of high-energy phosphate bonds for tissue metabolism. Suggested for use in cardiovascular disease and rheumatism but efficacy unproven.

Adrenaline
Sympathomimetic amine, alpha and beta adrenoceptor agonist. Produces vasoconstriction with rise in blood pressure, cardio-acceleration, broncho-dilatation. Used in bronchial asthma, acute allergic reactions, as peripheral vasoconstrictor and in cardiac arrest. Toxicity includes hypertension, pulmonary oedema and cardiac arrhythmias.

Agar
Purgative. Increases faecal bulk by same mechanism as **Methylcellulose** but less effective.

Alcofenac
Anti-inflammatory analgesic with actions similar to **Ibuprofen.**

Alcuronium
Skeletal muscle relaxant with uses and adverse effects similar to **Tubocurarine.**

Aldosterone
Naturally occurring adrenal (mineralocorticoid) steroid hormone. Acts mainly on salt and water metabolism by increasing salt retention in the kidney; has no useful anti-inflammatory activity. Used only in replacement therapy for adrenal insufficiency.

Allantoin
Used in creams and lotions to stimulate wound healing.

4

Allopurinol
Reduces formation of uric acid from purine precursors by inhibiting the enzyme xanthine oxidase. Used in primary and secondary gout.

Allylbarbituric aid
Barbiturate sedative with actions similar to **Amylobarbitone.**

Allylestrenol
Hormone with actions similar to **Progesterone.**

Aloes
Derived from species of aloe. Used as purgative, producing motion 6–12 hours after ingestion. Causes griping. Colours urine red. May cause nephritis in large doses.

Aloin
Extract of aloes: see **Aloes.**

Aloxiprin
Complex of **Aluminium antacids** and **Acetylsalicylic acid,** yielding these agents after breakdown in the gastro-intestinal tract.

Alphadolone
See **Alphaxalone.**

Alphaxalone
Steroid used as intravenous anaesthetic, in combination with **Alphadolone.**

Alprenolol
Beta adrenoceptor blocking drug with partial agonist activity (intrinsic sympathomimetic activity). Uses, side effects, etc. as for **Propranolol.**

Alum (Potassium aluminium sulphate)
Used as solid to stop bleeding; in powder for application to umbilical cord. Precipitates proteins and is a powerful astringent.

Aluminium antacids
Range of aluminium salts, used alone or complexed with other compounds. Neutralise gastric acid in treatment of peptic ulceration. Large doses cause constipation which may be prevented by combination with **Magnesium antacids.** May reduce absorption of other drugs, e.g. **Tetracyclines.**

Aluminium carbonate
Non-systemic (non-absorbable) antacid. Used in treatment of peptic ulceration where it produces longer neutralisation of acid than **Sodium bicarbonate.** Also used in prevention of urinary phosphate stones. May cause constipation. Reduces absorption of **Tetracyclines** given at same time.

Aluminium chlorohydrate
Used in antiperspirant preparations.

Aluminium glycinate
See **Aluminium antacids.**

Aluminium hydroxide
Non-systemic (non-absorbable) antacid. Neutralises gastric acid and binds phosphate ions in the gut. Used to treat peptic ulceration by reducing gastric acidity and also to increase phosphate excretion when phosphate retention is associated with stone formation (e.g. renal stones). Large doses cause constipation which may be prevented by combination with **Magnesium antacids.**

Alverine
Antispasmodic drug used in gut colic; related to **Papaverine.**

Amantadine
Antiviral agent for prophylaxis against influenza. Antiparkinsonian drug used in mild cases. Adverse effects include dry mouth, visual disturbance, confusion, hallucinations, ankle oedema.

Ambenonium
Anticholinesterase: similar to **Pyridostigmine.**

Ambucetamide
Antispasmodic used in preparations recommended for dysmenorrhoea.

Ambutonium
Anticholinergic with actions similar to **Atropine sulphate.** Used in treatment of peptic ulcer.

Amethocaine
Local anaesthetic similar to **Lignocaine.** Powerful surface activity but toxic by injection. Used topically in ophthalmology.

Amikacin
Antimicrobial with actions and uses similar to **Gentamycin.**

Amiloride
Potassium-sparing diuretic. Acts by inhibiting exchange of sodium for potassium in the distal tubule of the kidney. Relatively weak diuretic used when there is particular danger of potassium loss, e.g. fluid overload due to liver failure. Often combined with a thiazide (e.g. **Bendrofluazide**) or **Frusemide.** Danger of excessive potassium retention. May cause nausea, vomiting and diarrhoea.

Aminacrine
Skin disinfectant.

Aminoacridine
See **Aminacrine.**

Aminocaproic acid
Antifibrinolytic agent used to reverse effects of **Streptokinase** or other fibrinolytic activity.

Aminophylline (Theophylline ethylenediamine)
Relaxes smooth muscle, dilates bronchi, increases heart rate and force, has diuretic action. Used in cardiac and bronchial asthma. Given locally, intra-venously or by suppository. Adverse effects include nausea, vomiting if given orally; vertigo, restlessness, cardiac arrhythmias if given intravenously.

Amitriptyline
Antidepressant. Actions and adverse effects similar to **Imipramine.** Also has sedative and anxiolytic properties.

Ammonium acetate
Used as expectorant.

Ammonium chloride
Acidifying agent. Also used as expectorant and diuretic.

Ammonium polystyrine sulphonate
Ion exchange resin. Binds cations at pH 5 or above and preferentially binds sodium, releasing ammonium. Used to remove excess sodium, e.g. in cardiac failure. Large doses needed. May cause nausea, vomiting and constipation. If too much sodium is removed, excessive weight loss and weakness result.

Amoxycillin (c)
Similar to **Ampicillin** but better absorbed.

Amphetamine
Sympathomimetic amine. Increases heart rate and blood pressure by nor-adrenaline release from sympathetic nerve endings. Produces central stimula-tion through central noradrenergic and dopaminergic receptor activity. Many uses including as anorectic. Produces dependence and, in prolonged excessive use, psychosis. Now falling into disuse.

Amphomycin
Poorly absorbed antibiotic, used mainly in skin preparations for impetigo and infected dermatitis.

Amphotericin
Antifungal. Used topically for skin infections or by injection for severe generalised fungal infections. Topically it may produce irritation of skin. Systemic side effects are often severe, including headache, vomiting, fever, joint pains, convulsions and kidney damage.

Amphotericin B
Antibiotic active against a variety of fungal infections. Given intravenously for systemic infection. Adverse effects include fever, vomiting, nephrotoxicity.

Ampicillin
Penicillin antibiotic with broader spectrum of activity than benzylpenicillin; active against typhoid fever. Adverse effects as for **Benzylpenicillin.** Rash common if given to patients with infectious mononucleosis (glandular fever).

Amylmetacresol
Disinfectant used in mouth washes, gargles, lozenges. For mouth and throat infections.

Amylobarbitone
Barbiturate hypnotic/sedative. General depressant action on C.N.S. Used in treatment of insomnia and anxiety. Frequently has a 'hangover' effect with impairment of mental and physical performance. Tolerance and addiction may occur, with insomnia, delirium and convulsions on withdrawal. Metabolised by liver and therefore used with caution in liver disease. Induces its own metabolism and may thus affect other drugs with danger of adverse drug interaction. Coma with respiratory depression in overdosage. No antidote; treated by supportive measures.

Amylocaine
Local anaesthetic with actions similar to **Lignocaine.**

Ancrod
Anticoagulant enzyme from venom of Malayan pit viper. Given intravenously. May produce allergic reactions.

Anethole
Essential oil with odour of anise. Used as a carminative and expectorant.

Aneurine (Thiamine, Vitamin B₁)
Vitamin. Deficiency may cause cardiac failure (wet beri-beri), peripheral neuritis (dry beri-beri) and Wernicke's encephalopathy.

Angiotensin
Peptide pressor agent given by intravenous infusion in treatment of hypotensive states and shock. Adverse effects include headache and cardiac arrhythmias.

Antazoline
Antihistamine with actions and uses similar to **Promethazine.**

Antidiuretic hormone
See **Vasopressin.**

Aprotinin
Inhibits enzymes that digest proteins. Used in acute pancreatitis. Adverse effects include allergic reactions.

Arginine
Essential amino acid used in treatment of liver coma and in tests of growth hormone secretion.

Ascorbic acid
See **Vitamin C.**

Aspirin
See **Acetylsalicylic acid.**

Atenolol
Beta adrenoceptor blocking drug with limited cardioselectivity. Uses, side effects, etc. as for **Propranolol.**

Atropine
See **Atropine sulphate.**

Atropine methonitrate
Anticholinergic with actions, uses and adverse effects similar to **Atropine sulphate** but has less effects upon C.N.S. and is sometimes considered less toxic.

Atropine sulphate
Parasympatholytic derivative of belladonna plants, e.g. deadly nightshade. Blocks peripheral autonomic cholinergic nerve junctions. Causes dilation of pupils, paralysis of ocular accommodation, tachycardia, reduced gut motility, decreased secretions and C.N.S. stimulation. Used intravenously in treatment of bradycardia and anticholinesterase poisoning (e.g. due to organophosphorus insecticides), intramuscularly as part of pre-operative medication, and topically in the eye for optical refraction in children. For Parkinsonism and peptic ulceration it has largely been replaced by other anticholinergics. May cause dry mouth, blurred vision, glaucoma and retention of urine. In overdosage there is tachycardia, fever, flushed skin, dehydration and excitement. Anticholinesterase, e.g. **Neostigmine,** may be used as antidote.

Attapulgite
Form of magnesium aluminium silicate used as absorbent and adsorbent agent in treatment of acute poisoning, in diarrhoea and in topical deodorant preparations.

Aurothiomalate sodium
Preparation of gold for intramuscular injection in treatment of active rheumatoid arthritis. Adverse effects include allergic reactions such as rashes, blood dyscrasias, jaundice, kidney dysfunction, peripheral neuritis and encephalitis.

Azapropazone
Non-steroid anti-inflammatory analgesic used in arthritic conditions. Adverse effects include some gastro-intestinal disturbances, although not as often as with **Acetylsalicylic acid,** and allergic rashes.

Azathioprine
Derivative of **Mercaptopurine,** used primarily as immunosuppressant agent in patients receiving organ transplants. Adverse effects include bone marrow depression.

B

Bacitracin
Peptide antibiotic active mainly on gram-positive cocci. Nephrotoxic on systemic administration. Only used topically for skin infections.

Baclofen
Used in treatment of spasticity, mode of action uncertain. Adverse effects include nausea and sedation.

Bamethan
Vasodilator with actions and adverse effects similar to **Tolazoline.**

Bamifylline
Bronchodilator with actions similar to **Aminophylline.**

Beclamide
Anticonvulsant drug that is claimed to possess stimulant rather than sedative properties.

Beclomethasone
Potent synthetic corticosteroid similar to **Dexamethasone.** Used by inhalation for treatment of asthma.

Belladonna extract
Plant extract containing **Atropine sulphate** and having similar actions, uses and adverse effects.

Bemegride
C.N.S. stimulant formerly used as respiratory stimulant in barbiturate over-dosage. Adverse effects include convulsions and delirium.

Benapryzine
Antiparkinsonian drug with actions similar to **Benzhexol.**

Bendrofluazide
Thiazide diuretic used in the treatment of fluid overload and in control of high blood pressure. Acts by reducing sodium reabsorption in the kidney. Less potent than **Frusemide** and **Mersalyl** but has longer action. It is effective orally. May cause excessive loss of potassium in urine and increase in blood uric acid or glucose. Consequently can produce symptoms of hypokalaemia, gout or diabetes.

11

Benethamine penicillin
Long-acting form of **Benzylpenicillin,** with similar actions and adverse effects.

Benorylate
Analgesic anti-inflammatory combination that breaks down in the body into **Acetylsalicylic acid** and **Paracetamol,** and has the actions of each.

Benperidol
Tranquilliser with actions, uses, etc. similar to **Haloperidol.**

Benserazide
Used with **Levodopa** in Parkinson's disease. Prevents peripheral breakdown of levodopa, allowing reduced dosage and decreased side effects.

Benzalkonium
Topical disinfectant used in creams, lozenges, irrigating solutions.

Benzamine
Local anaesthetic with actions similar to **Lignocaine.**

Benzathine penicillin
Long-acting form of **Benzylpenicillin,** with similar actions and adverse effects.

Benzethonium
Similar to **Benzalkonium.**

Benzhexol
Antispasmodic parasympatholytic used in Parkinsonism of all causes. May produce dry mouth, blurred vision, constipation, hesitancy of micturition, confusion, hallucinations. Contra-indicated in glaucoma, prostate hypertrophy. In overdosage dry mouth, nausea, vomiting, excitement, confusion, hot dry skin, rapid pulse, fixed dilated pupils. Depression of respiration and hypotension with loss of consciousness in late stages. **Physostigmine** salicylate is effective antidote.

Benzilonium
Parasympatholytic with peripheral effects, toxic effects, etc. similar to **Atropine.** Used as antispasmodic for gastro-intestinal disorders and to reduce gastric acid secretion in peptic ulceration.

Benzocaine
Weak local anaesthetic similar to **Lignocaine.** Used in proprietary preparations for sore throats.

Benzoctamine
Anxiolytic. Actions uses and adverse effects similar to **Diazepam.**

Benzoic acid
Used topically for mild fungus infections of skin.

Benzoyl peroxide
Powder used in dusting powders, and in creams and lotions in treatment of acne.

Benzthiazide
Diuretic with actions similar to **Chlorothiazide.**

Benztropine
Antispasmodic parasympatholytic used in Parkinsonism. Similar to **Benzhexol** but more potent and can be given by intramuscular injection. Particularly useful in treating drug-induced Parkinsonism.

Benzyl benzoate
Used as insect repellent, in treatment of scabies and as antipruritic. May cause allergic rashes.

Benzyl nicotinate
Vasodilator related to **Nicotinic acid.**

Benzylpenicillin
Bactericidal antibiotic (see **Penicillin**). Unstable at acid pH, poorly active by mouth. Given parenterally. Active against most gram-positive and some gram-negative organisms. Inactivated by penicillinase. Adverse effects include hypersensitivity reactions, both immediate and delayed, and encephalopathy with convulsions if given intrathecally or in massive doses.

Bephenium
Used in treatment of hookworms. Adverse effects include nausea, vomiting, vertigo.

Betahistine
Vasodilator with actions similar to **Histamine.** Used in Ménière's disease to reduce episodes of dizziness.

Betamethasone
Potent synthetic corticosteroid similar to **Dexamethasone.**

Betazole
Related to **Histamine** with similar actions and uses.

Bethanechol
Parasympathomimetic drug with actions of **Acetylcholine.**

Bethanidine
Adrenergic neurone blocking drug. Used in hypertension. Adverse effects as for **Guanethidine.**

13

Bile salts
Extracted from animal bile. Used to stimulate bile flow without increasing its contents of bile salts and pigment, e.g. after biliary operations. Included in some compound preparations for treatment of biliary insufficiency but of doubtful efficacy.

Biperiden
Parasympatholytic antispasmodic used in treatment of Parkinsonism of all causes. Similar actions, adverse effects and overdose effects to **Benzhexol**.

Bisacodyl
Purgative that acts by stimulating sensory nerve endings in wall of large bowel. Available for oral and rectal use. Suppositories may cause mild burning sensation in the rectum.

Bismuth aluminate
See **Bismuth antacids**.

Bismuth antacids
Insoluble bismuth salts have weak antacid properties and are claimed to protect the stomach. Largely superseded by more effective antacids. Prolonged, excessive use may allow sufficient absorption to cause toxicity with kidney damage, liver damage and C.N.S. effects.

Bismuth carbonate
See **Bismuth antacids**.

Bismuth-formic-iodide
Topical anti-infective powder.

Bismuth oxide
See **Bismuth antacids**.

Bismuth subgallate
Insoluble powder used for eczema and as suppositories for haemorrhoids.

Bismuth subnitrate
See **Bismuth antacids**.

Bithionol
Topical anti-infective seldom used because of photosensitisation.

Borax
Similar actions to **Boric acid**.

Boric acid
Weak anti-infective powder used in dusting powders, lotions and ointments.

Bran
Purgative, non-irritant. By-product of milling of wheat. Contains indigestible cellulose which increases intestinal bulk. Crude bran is unpalatable; processed bran is pleasant cereal. Large doses needed for effect. Danger of bowel obstruction if pre-existing bowel narrowing.

Bretylium
Adrenergic neurone blocking drug with actions similar to **Guanethidine.** Used mainly in cardiac arrhythmias. Side effects have limited its use as antihypertensive.

Bromhexine
Mucolytic, expectorant. Administered orally. Said to increase secretion of fluid by respiratory tract and to break down mucus. Used when viscid mucus is troublesome, e.g. chronic bronchitis. May cause gastro-intestinal side effects. Not to be given if there is pre-existing peptic ulceration.

Bromides
C.N.S. depressants, now largely superseded by safer drugs.

Bromocriptine
Stimulates dopamine receptors. Used in treatment of acromegaly, for inhibition or suppression of lactation, and in conditions due to excessive prolactin secretion including some cases of infertility. Adverse effects include nausea, hypotension, cold extremities.

Brompheniramine
Antihistamine with actions similar to **Promethazine.**

Bromvaletone
Similar actions to **Carbromal.**

Bronopol
Antibacterial preservative used in topical preparations.

Buclizine
Antihistamine/anti-emetic drug with actions similar to **Promethazine.**

Buclosamide
Antifungal agent for topical application.

Bufexamac
Analgesic/anti-inflammatory with actions similar to **Indomethacin.**

Bufylline
Bronchodilator with actions similar to **Aminophylline.**

Bum

Bumetanide
Potent diuretic with actions and uses similar to **Frusemide**. Adverse effects similar to **Bendrofluazide**. May also cause transient deafness.

Bupivacaine
Local anaesthetic similar to **Lignocaine** but produces longer anaesthesia.

Busulphan
Cytotoxic drug used in neoplastic disease, particularly myeloid leukaemia. Adverse effects include skin pigmentation, cataract, pulmonary fibrosis, bone marrow depression.

Butacaine
Local anaesthetic used by injection or by spray onto the mucosa of the nose and throat. Actions and adverse effects similar to **Lignocaine**.

Butethamate
Sympathomimetic amine with actions similar to **Ephedrine**.

Butobarbitone
Barbiturate hypnotic essentially like **Amylobarbitone**.

Butoxyethyl nicotinate
See **Nicotinic acid**.

Butriptyline
Antidepressant with actions and uses similar to **Imipramine**.

Butyl aminobenzoate
Local anaesthetic for topical use.

C

Caffeine
Active principle from tea and coffee, used as mild C.N.S. stimulant. Adverse effects include ˙restlessness, excitement and dependence after prolonged excessive ingestion.

Calamine
Zinc carbonate used in dusting-powders, creams, lotions, etc.

Calciferol
See **Vitamin D.**

Calcitonin
Hormone from parathyroid glands, involved in control of calcium metabolism. Overdosage causes abnormally high levels of calcium in blood.

Calcium carbimide
Actions and uses similar to **Disulfiram,** but produces fewer adverse effects.

Calcium carbonate
Non-systemic (non-absorbable) antacid. Used in treatment in peptic ulceration where it produces longer neutralisation of acid than **Sodium bicarbonate.** Frequent use may cause constipation. Small amounts are absorbed and in some subjects may cause renal stones (nephrocalcinosis).

Calcium gluconate
Source of calcium for deficiency states.

Calcium iodide
Used as expectorant.

Calcium lactate
See **Calcium gluconate.**

Calcium polystyrene sulphonate
Ion exchange resin used to treat electrolyte abnormalities by changing absorption or excretion in the gut.

Calcium sulphaloxate
Sulphonamide antibacterial with actions similar to **Sulphadimidine.** Very little absorbed. Used for prevention or treatment of mild infective diarrhoea.

17

Camphene
See **Camphor.**

Camphor
Used internally as carminative and externally as rubefacient.

Candicidin
Antifungal antibiotic used locally for vaginal and skin infections.

Capreomycin
Peptide antibiotic, mainly used in tuberculosis. Adverse effects include ototoxicity and nephrotoxicity.

Capsicum
Essential oil used internally as carminative and as externally as rubefacient.

Caraway
Essential oil used as carminative.

Carbachol
Parasympathomimetic with actions and adverse effects similar to **Acetylcholine** but more prolonged. Used as miotic eye drops and for improvement
of post-operative intestinal or bladder muscle tone.

Carbamazepine
Anticonvulsant. Acts by suppressing epileptic discharges in the brain. Used
in prevention of epilepsy and in suppression of pain in trigeminal neuralgia
but is not an analgesic. May cause drowsiness, blurred vision, dizziness and
gastro-intestinal upsets. Skin rashes and adverse effects on the liver and bone
marrow are relatively common. Coma with convulsions in overdosage. No
antidote; supportive treatment only.

Carbaryl
Cholinesterase inhibitor. Used topically as an insecticide, e.g. for lice.

Carbenicillin
Penicillin antibiotic, particularly active against gram-negative bacteria
especially *Pseudomonas* and *Proteus*. Adverse effects as for **Benzylpenicillin.**

Carbenoxolone
Used in treatment of gastric and duodenal ulcers and for mouth ulcers.
Adverse effects include oedema, hypertension, hypokalaemia, muscle pain.

Carbidopa
Similar actions to **Benserazide.**

Carbimazole
Depresses formation of thyroid hormone. Used in treatment of hyperthyroidism. Adverse effects include allergic rashes, nausea, diarrhoea, blood
abnormalities.

Carbinoxamine
Antihistamine with actions similar to **Promethazine.**

Carbocisteine
Used to reduce viscosity of sputum.

Carbromal
Weak hypnotic. May produce dependence. Adverse effects include rashes and purpura. Chronic effects (bromism) include mental depression and slurring of speech. Acute intoxication produces respiratory failure.

Carfecillin
Similar to **Carbenicillin.**

Carisprodol
Used to treat painful muscle spasm.

Cascara
Purgative from bark of buckthorn tree. Stimulates gut movement via the nerve plexus in the large bowel wall. Produces reddish-brown discoloration of urine and may cause excessive catharsis. Excreted in milk of lactating mothers and may cause diarrhoea in infants. Prolonged use causes black pigmentation in colon (melanosis coli).

Castor oil
Purgative with action upon small intestine as well as large intestine, useful when prompt evacuation is required, e.g. before bowel X-rays. Chronic use not recommended as it causes reduced absorption of nutrients.

Centella asiatica
Plant extract used in skin preparations to promote healing.

Cephaeline
Used in amoebiasis. Similar to **Emetine.**

Cephalexin
Cephalosporin antibiotic with similar activity and adverse effects to **Cephalothin** but well absorbed by mouth.

Cephaloridine
Cephalosporin antibiotic, administered parenterally. May cause renal damage, particularly if given with **Frusemide.** Other adverse effects include hypersensitivity reactions.

Cephalosporins
Bactericidal antibiotics that inhibit bacterial cell wall synthesis. Have similar basic structure to **Penicillins,** but are relatively resistant to penicillinase.

Cephalothin
Cephalosporin antibiotic, particularly useful against penicillinase-producing *Staphylococcus aureus*. Must be given parenterally. Adverse effects mainly hypersensitivity reactions.

Cephazolin
Antibiotic similar to **Cephalexin.**

Cephradine
Cephalosporin antibiotic with actions similar to **Cephalexin.**

Cetalkonium
Topical disinfectant.

Cetrimide
Topical disinfectant used in many skin preparations.

Cetyl alcohol
Used in manufacture of ointments and creams.

Cetylpyridinium
Topical disinfectant used in skin and mouth preparations.

Charcoal
Used as adsorbent in first aid treatment of poisoning by drugs and toxins. Also used to treat diarrhoea.

Chloral hydrate
Hypnotic. Available only as a liquid. Converted by liver to trichloroethanol which causes generalised C.N.S. depression. Used for insomnia, especially in children and the elderly. Relatively 'safe'. Addiction is rare. Interacts with oral anticoagulants increasing their effect and rate of elimination. Coma in overdosage. No antidote; treated by supportive measures.

Chlorambucil
Cytotoxic drug related to **Mustine hydrochloride.** Used in neoplastic conditions of lymphoid tissues. Adverse effects include bone marrow depression.

Chloramphenicol
Bacteriostatic antibiotic, broad spectrum, which should be reserved for treatment of typhoid fever and life-threatening infections. Adverse effects include aplastic anaemia. Produces 'grey baby syndrome' in neonates and premature babies.

Chlorbutol
Antibacterial and antifungal preservative for topical applications.

Chlorcyclizine
Antihistamine with similar actions and adverse effects to **Promethazine**. Used mainly as an anti-emetic.

Chlordantoin
Topical antifungal agent.

Chlordiazepoxide
Benzodiazepine anxiolytic similar to **Diazepam** but less hypnotic and less anticonvulsant activity. Used in treatment of anxiety.

Chlorexolone
Diuretic with actions, uses and adverse effects similar to **Bendrofluazide**.

Chlorhexidine
Topical disinfectant used in skin preparations and as preservative in eye drops.

Chlormethiazole
Sedative/hypnotic/anticonvulsant. Depressant action on C.N.S. Used for sedation or hypnosis in agitated or confused patients especially the elderly. Also for treatment of acute withdrawal symptoms in alcoholics and drug addicts and control of sustained epileptic fits (status epilepticus). May cause tingling in nose and sneezing. Effects potentiated by phenothiazines (e.g. **Chorpromazine**) and **Haloperidol**. Coma with respiratory depression in overdosage. No antidote. Symptomatic treatment is adequate.

Chlormezanone
Minor tranquilliser essentially similar to **Meprobamate**.

Chlorocresol
Disinfectant used in sterilising solutions and as a preservative in creams.

Chlorofluoromethane
Aerosol propellant for dugs administered by inhalation. Also used as a spray for muscle pain where it produces local anaesthesia due to intense coldness.

Chlorophenoxyethanol
Topical antibacterial.

Chloroquine
Antimalarial agent, which has also been used in rheumatoid arthritis. Adverse effects include skin pigmentation, alopecia, neuropathy, corneal, retinal damage.

Chlorothiazide
Thiazide diuretic similar to **Bendrofluazide**.

Chlorothymol
Topical antiseptic.

Chlorotrianisene
Synthetic female sex hormone, used in menopausal symptoms and to suppress lactation. Adverse effects similar to **Oestradiol**.

Chloroxylenol
Topical disinfectant used chiefly on skin.

Chlorphenesin
Topical antibacterial/antifungal agent.

Chlorpheniramine
Antihistamine with actions, uses and adverse effects similar to **Promethazine**.

Chlorphenoxamine
Antihistamine similar to **Promethazine** used in Parkinson's disease.

Chlorphentermine (c)
Anorectic, sympathomimetic amine. Actions and adverse effects similar to **Diethylpropion**.

Chlorpromazine
Phenothiazine tranquilliser. Causes selective depression of the brain structures responsible for control of behaviour and wakefulness. Has anticholinergic and alpha adrenergic blocking effects amongst other pharmacological effects. Used in psychotic disorders, particularly schizophrenia and agitated depression; in terminal illness to enhance analgesia; to control nausea and vomiting; and for hiccups. Adverse effects include postural hypotension, dry mouth, blurred vision, extrapyramidal syndromes, cholestatic jaundice, photosensitivity and deposits in lens and cornea. Used only with caution in liver disease and epilepsy (may precipitate convulsions). In overdosage causes coma, extrapyramidal signs, convulsions, hypotension and arrhythmias. No antidote. Supportive treatment only.

Chlorpropamide
Oral antidiabetic drug that stimulates pancreatic insulin release in maturity-onset diabetes mellitus. Adverse effects include hypoglycaemia, allergic reactions, jaundice, flushing with alcohol. Action may be potentiated by salicylates and sulphonamides. Sometimes used in diabetes insipidus.

Chlorprothixene
Phenothiazine tranquilliser essentially similar to **Chlorpromazine**.

Chlorquinaldol
Topical antibacterial/antifungal used in skin infections.

Chlortetracycline
Bacteriostatic antibiotic with actions, adverse effects and interactions similar to **Tetracycline.**

Chlorthalidone
Diuretic essentially similar to **Bendrofluazide.**

Cholesterol
Natural fatty constituent of all animal cells. Used topically in creams for soothing and water-absorbing properties.

Cholestyramine
Resin that binds bile salts in gut. Used in pruritus associated with jaundice, and to reduce blood cholesterol. Adverse effects include nausea, diarrhoea constipation.

Choline salicylate
Similar actions to **Acetylsalicylic acid.**

Choline theophyllinate
Oral preparation of **Theophylline** with actions similar to **Aminophylline.** Main use is in chronic bronchitis.

Chymotrypsin
Animal pancreatic enzyme used to reduce soft tissue inflammation, particularly associated with trauma. Adverse effects include allergic reactions.

Cimetidine
Selectively blocks histamine receptors mediating gastric acid secretion. Used in peptic ulceration and gastric hyperacidity.

Cinchocaine
Local anaesthetic with actions similar to **Lignocaine.**

Cinnarizine
Antihistamine similar to **Promethazine,** chiefly used in treatment of vertigo and vomiting.

Clemastine
Antihistamine, with actions and uses similar to **Promethazine** but with less sedative effects.

Clemizole
Antihistamine similar to **Promethazine.**

Clidinium
Actions similar to **Atropine.** Used in treatment of peptic ulcer and gastric hyperacidity.

Clindamycin
Antibiotic with actions and adverse effects similar to **Lincomycin** but better absorbed.

Clioquinol
Used in treatment of gut amoebiasis and to protect against gut infections, topically for skin infections. Prolonged large oral doses may produce neuropathy.

Clobetasol
Topical **Corticosteroid** for psoriasis and eczema.

Clobetasone
Topical **Corticosteroid** for psoriasis and eczema.

Clofazimine
Antileprotic, anti-inflammatory drug, used for control of reactions occurring with **Dapsone** treatment. Adverse effects include skin pigmentation, red urine, diarrhoea.

Clofibrate
Reduces blood cholesterol and fats. Used in patients with raised levels of these constituents. Adverse effects include nausea, diarrhoea, muscle pain, weakness. Potentiates anticoagulants.

Clomipramine
Antidepressant drug with actions and uses similar to **Imipramine.**

Clomocycline
Bacteriostatic antibiotic with actions, adverse effects and interactions similar to **Tetracycline.**

Clonazepam
Benzodiazepine anticonvulsant similar to **Diazepam** but has greater anticonvulsant activity. Used intravenously for control of status epilepticus and orally for prevention of all types of epilsepsy.

Clonidine
Reduces sympathetic activity by central action, and reduces vascular reactivity. Used in hypertension and in migraine. Antihypertensive effect blocked by tricyclic antidepressants. Adverse effects include sedation, depression, dryness of mouth, fluid retention. Rapid withdrawal may be associated with 'rebound hypertension'.

Clopamide
Diuretic essentially similar to **Bendrofluazide.**

Clorexolone
Diuretic essentially similar to **Bendrofluazide.**

Clorprenaline
Bronchodilator similar to **Ephedrine.**

Clotrimazole
Antifungal agent used topically for skin infections with *Candida.*

Cloxacillin
Penicillinase-resistant **Penicillin** with actions and adverse effects similar to **Benzylpenicillin.** Use restricted to treatment of penicillinase-producing *Staphylococcus aureus* infections.

Coal tar
Used in topical preparations for eczema and psoriasis.

Cobalt tetracemate (cobalt edetate)
Antidote for cyanide poisoning. Binds with cyanide and prevents its effects upon cell metabolism.

Cocaine (c)
Local anaesthetic. Stabilises nerve cell membranes to prevent impulse transmission. Little used except topically in eye or respiratory passages. Frequent use may cause corneal ulceration. Stimulates C.N.S. with euphoria and consequent risk of addiction. Chronic misuse leads to delusions, hallucinations and paranoia.

Codeine
Weak narcotic analgesic. Used for somatic (deep) pain often combined with **Aspirin** or **Paracetamol.** Also causes constipation and suppresses the cough reflex. May therefore be used as an antidiarrhoeal and in cough mixtures. Addiction very unusual. Coma with respiratory depression in overdosage. **Naloxone** is antidote.

Colaspase
Cytotoxic drug: see **L-Asparaginase.**

Colchicine
Used for relief of pain in acute gout. Adverse effects include nausea, vomiting, colicky pain, diarrhoea.

Colistin
Antibiotic: see **Polymyxin.**

Colophony
Resin used in protective topical preparations.

Copper sulphate
Used as an emetic, together with iron in treatment of anaemia, and as astringent in topical preparations. Large doses may cause copper poisoning. Syrup of **Ipecacuanha** is generally considered a safer emetic.

2

Corticosteroids
General term to include natural and synthetic steroids with actions similar to **Hydrocortisone**, which is produced in the adrenal cortex. They possess anti-inflammatory and salt-retaining properties. Adverse effects include oedema, hypertension, diabetes, bone thinning with fractures, muscle wasting, infections, psychosis.

Corticotrophin
Pituitary hormone that controls functions of adrenal cortex.

Cortisone
Naturally occurring adrenal (glucocorticoid) steroid hormone. Has effects upon fat, protein and carbohydrate metabolism and possesses marked anti-inflammatory activity. Used for replacement therapy in adrenal insufficiency, anti-inflammatory activity in a wide range of conditions, and immuno-suppression after organ transplantation or in certain leukaemias. Adverse effects include retention of salt and water, fulminating infections, osteo-porosis, peptic ulceration, muscle wasting, hypertension, diabetes mellitus, weight gain, moon face, cataracts and psychiatric disturbance. On withdrawal of large doses after long periods of treatment there may be failure of the natural adrenal hormone secretion.

Cotrimoxazole
Antimicrobial. Combination of **Sulphamethoxazole** and **Trimethoprim**. Broad antibacterial spectrum, active against typhoid fever. Adverse effects include rashes and blood dyscrasias.

Cresol
Antiseptic. Used as disinfectant or preservative and also as an inhalant for relief of congestion in bronchitis, asthma and the common cold. If ingested in concentrated solutions there may be local corrosion, depression of the C.N.S. and damage to the liver and kidneys.

Cropropamide
C.N.S. stimulant used to stimulate respiration. Adverse effects include tremor and restlessness.

Crotamiton
Topical treatment for scabies.

Crotethamide
As for **Cropropamide**.

Cyanocobalamin
See **Hydroxocobalamin**.

Cyclandelate
Peripheral vasodilator acting by relaxation of muscle in blood vessel walls. Used in treatment of peripheral vascular insufficiency, e.g. Raynaud's syndrome. May cause dizziness, flushing, headache and nausea.

26

Cyclizine
Antihistamine with actions similar to **Promethazine**. Main use as anti-emetic.

Cyclobarbitone
Barbiturate hypnotic with actions, uses and adverse effects similar to **Amylobarbitone.**

Cyclopentamine
Sympathomimetic amine. Actions and uses similar to **Ephendrine.**

Cyclopenthiazide
Thiazide diuretic similar to **Bendrofluazide.**

Cyclopentolate
Anticholinergic with actions and adverse effects similar to **Atropine** but with more rapid onset and shorter duration. Used as eye drops to dilate the pupil and to assist optical refraction.

Cyclophosphamide
Cytotoxic drug used in neoplastic disease. Adverse effects include alopecia, cystitis, bone marrow depression.

Cycloserine
Antibiotic used in tuberculosis and in *Escherichia coli* and *Proteus* infections. Adverse effects include ataxia, drowsiness, convulsions.

Cycrimine
Parasympatholytic used in treatment of Parkinsonism. Similar actions, etc. to **Benzhexol.**

Cyproheptadine
Antihistamine similar to **Promethazine.** Stimulates appetite.

Cyproterone
Used in excessive hirsutism and abnormal sexual activity.

Cysteamine
Antidote for severe poisoning due to **paracetamol** where it is thought to prevent liver damage by reducing the concentration of a toxic metabolite of paracetamol. Must be given by intravenous injection and may cause marked adverse effects including protracted nausea and vomiting. Must not be given more than 10 hours after the overdose as it may exacerbate liver damage.

Cytarabine (Cytosine arabinoside)
Antiviral agent used systemically for herpes encephalitis. Cytotoxic, used in treatment of leukaemia and Hodgkin's disease. Adverse effects include bone marrow depression.

Cytosine arabinoside
See **Cytarabine.**

D

Dacarbazine
Cytotoxic. May cause bone marrow suppression.

Dakin's solution
Contains calcium hypochlorite, **Sodium bicarbonate, Boric acid.** Used as wound disinfectant.

Danazol
Used in endocrine disturbances where pituitary control of gonad hormone production is required.

Danthron
Purgative with actions etc. similar to **Cascara.**

Dantrolene
Used in control of skeletal muscle spasticity. Adverse effects include sedation, weakness, diarrhoea.

Dapsone
Sulphone drug used in treatment of leprosy. Adverse effects include allergic dermatitis, nausea, vomiting, tachycardia, haemolytic anaemia, liver damage.

Daunomycin
See **Daunorubicin.**

Daunorubicin (Rubidomycin, Daunomycin)
Cytotoxic antibiotic used in neoplastic disease. Adverse effects include cardiotoxicity and bone marrow depression.

Debrisoquine
Adrenergic neurone-blocking drug, used in hypertension. Adverse effects as for **Guanethidine.**

Deglycyrrhizinised liquorice
Mild anti-inflammatory agent. Used in treatment of peptic ulcer. Adverse effects include oedema and hypertension.

Dehydrocholic acid
Used to stimulate secretion of bile flow without increasing its content of bile solids, e.g. after surgery of biliary tract.

Demecarium
Anticholinesterase used by instillation into eye in glaucoma. Actions those of **Acetylcholine.**

Demeclocycline
Antibiotic with actions similar to **Tetracycline.**

Demethylchlortetracycline
Bacteriostatic antibiotic with actions, adverse effects and interactions similar to **Tetracycline.**

Deoxycortone
Potent salt-retaining corticosteroid used by injection, implant or sublingually in adrenal insufficiency. Adverse effects as for **Corticosteroids.**

Deptropine
Antihistamine similar to **Promethazine.**

Dequalinium
Topical antibacterial/antifungal used in oral infections.

Deserpidine
See **Reserpine.**

Desferrioxamine
Binds with iron. Used orally and parenterally in treatment of acute iron poisoning and in conditions associated with excessive iron storage in tissues, where it increases urinary iron excretion. Adverse effects include allergic reactions.

Desipramine
Antidepressant. Active metabolite of **Imipramine,** whose actions and adverse effects it shares.

Desmopressin
Synthetic form of **Vasopressin** for use nasally in diabetes insipidus.

Desonide
Topical corticosteroid for psoriasis and eczema.

Desoxyribonuclease
Animal pancreatic enzyme used to resolve clots and exudates associated with trauma and inflammation.

Dexamethasone
Potent synthetic corticosteroid with actions, etc. similar to **Cortisone.** Anti-inflammatory activity is much increased in potency with no increase in salt- and water-retaining activity.

Dexamphetamine (c)
See **Amphetamine.**

Dextrans
Polysaccharides used intravenously instead of blood or plasma to maintain blood volume and assist capillary flow. Adverse effects include allergic reactions.

Dextromethorphan
Cough suppressant. Adverse effects include slight psychic dependence and abuse.

Dextromoramide
Narcotic analgesic essentially similar to **Morphine** but more reliable when taken by mouth. Useful in the management of severe chronic pain in terminal disease.

Dextropropoxyphene
Weak narcotic analgesic with potency less than **Codeine.** Used in moderate pain, commonly with **Paracetamol** when the latter drug is not fully effective. In normal doses causes less nausea, vomiting and constipation than codeine. Coma with depressed respiration in overdosage. **Naloxone** is antagonist.

Dextrothyroxine
See: D-**Thyroxine.**

Diamidino-diphenylamine
Topical antiseptic.

Diamorphine (Heroin) (c)
Narcotic analgesic similar to **Morphine.** Less likely to cause nausea, vomiting, constipation and hypotension but greater euphorant action makes it more addicting and liable to greater abuse.

Diamthazole
Topical antifungal agent. Adverse effects include convulsions if absorbed.

Diazepam
Benzodiazepine minor tranquilliser (anxiolytic)/hypnotic with anticonvulsant properties. Acts centrally on the limbic system. Used in treatment of anxiety and as a hypnotic. Useful also in reduction of muscle tone in spasticity and as an anticonvulsant given intravenously for status epilepticus. May cause ataxia, nystagmus and sedation. Coma in overdosage but little respiratory depression. No antidote. Supportive treatment is adequate.

Diazoxide
Used to reduce blood pressure in severe hypertension and to increase blood sugar level in hypoglycaemia. Adverse effects include excessive hair growth, nausea, vomiting, oedema, diabetes, hypotension.

Dibenzepin
Antidepressant with actions similar to **Imipramine.**

Dibromopropamide
Topical antibacterial/antifungal.

Dichloralphenazone
Hypnotic. Combination of **Chloral hydrate** and **Phenazone.** Available as tablets and elixir. Converted back to parent compounds by the liver. Used for insomnia especially in children and the elderly. Relatively 'safe'. Addiction is rare but rashes and blood disorders may be caused by phenazone.

Dichlorobenzyl alcohol
Topical antiseptic used in oral preparations.

Dichlorofluoromethane
See **Chlorofluoromethane.**

Dichlorophen
Used in treatment of tapeworms. Adverse effects include nausea, vomiting, bowel colic.

Dichlorphenamide
Used in treatment of respiratory failure from chronic bronchitis, and in glaucoma. Adverse effects include electrolyte imbalance.

Dicophane
Insecticide used as dusting powder and lotion for fleas and lice. Very toxic if absorbed.

Dicoumard
Anticoagulant with actions, interactions and adverse effects similar to **Warfarin.**

Dicyclomine
Parasympatholytic used in spasm of gastro-intestinal and urinary tracts and to reduce gastric acid in peptic ulceration. Actions, etc. similar to **Atropine** but weaker.

Dienoestrol
Synthetic female sex hormone used for menopausal symptoms and for suppressing lactation. Adverse effects include nausea, vaginal bleeding, oedema.

Diethylamine salicylate
Rubefacient with actions similar to **Salicylic acid.**

Diethylcarbamazine
Used in filariasis. Adverse effects include anorexia, nausea, vomiting. Allergic reactions may accompany release of foreign proteins on death of the worms.

Diethylpropion
Anorectic, sympathomimetic amine. Actions those of **Amphetamine** but less central stimulation and abuse potential

Diflucortolone
Corticosteroid for topical use in inflammatory skin conditions. Actions and adverse effects similar to **Cortisone.**

Digitalis
Crude foxglove extract with same actions, etc. as **Digoxin** but content of active drug is less reliable.

Digitoxin
Foxglove derivative with similar actions, uses, etc. to **Digoxin.**

Digoxin
Foxglove derivative. Increases force of contraction of heart and slows heart rate, thus making cardiac function more efficient. Used in heart failure and certain abnormal heart rhythms. Influenced by serum potassium levels and by kidney function. In therapeutic overdose causes vomiting, abdominal pain, diarrhoea, impaired colour vision, slow heart rate, abnormal heart rhythms.

Dihydrocodeine
Mild narcotic analgesic. Similar to **Codeine** but more potent in relief of pain and more likely to cause constipation. (c) if given by injection.

Dihydroergocornine
See **Dihydroergotoxine.**

Dihydroergocristine
See **Dihydroergotoxine.**

Dihydroergokryptine
See **Dihydroergotoxine.**

Dihydroergotamine
For migraine. Drops, tablets or intramuscular injection. Used both for prevention and for symptomatic treatment. Has vasoconstrictor effects similar to **Ergotamine** but milder and with much reduced tendency to hypertension or effects on the uterus. No evidence of ergotism on prolonged or excessive use.

Dihydroergotoxine
Mixture of **Dihydroergocornine, Dihydroergocristine** and **Dihydroergokryptine,** ergot derivatives that are alpha adrenoceptor blockers and vasodilators, used in peripheral and cerebral vascular disease. Adverse effects include nausea and nasal stuffiness.

Dihydrostreptomycin
Antibiotic with actions similar to **Streptomycin** but more toxic to hearing.

Dihydrotachysterol
Closely related to **Vitamin D** and has similar actions.

Di-iodohydroxyquinoline
Used orally for amoebiasis and topically as skin antiseptic.

Dimenhydrinate
Antihistamine/anti-emetic with actions similar to **Promethazine.**

Dimercaprol
Binds with heavy metals. Used parenterally in treatment of heavy metal poisoning to increase urinary metal excretion. Adverse effects include nausea, vomiting, hypertension.

Dimethicone
Silicone used in protective creams and in antacid preparations.

Dimethindene
Antihistamine similar to **Promethazine.**

Dimethisoquin
Topical local anaesthetic used in lotions and ointments. Adverse effects include allergy and eye irritation.

Dimethothiazine
Antihistamine with actions similar to **Promethazine.**

Dioctyl sodium sulphosuccinate
Purgative. Lowers surface tension of faecal mass allowing water to penetrate and soften faecal matter. Should not be given together with mineral oil laxatives (e.g. **Liquid paraffin**) as this drug may enhance absorption of the oil.

Diphenhydramine
Antihistamine drug with actions similar to **Promethazine.**

Diphenoxylate
Reduces gut motility. Used in control of diarrhoea. Related to **Morphine;** adverse effects include drowsiness, euphoria, respiratory depression, coma, dependence.

Diphenylpyraline
Antihistamine similar to **Promethazine.**

Dipipanone (c)
Narcotic analgesic essentially similar to **Methadone.**

Diprophylline
Bronchodilator with actions similar to **Aminophylline.**

Dipyridamole
Used in treatment of angina. Reduces platelet stickiness. Adverse effects include flushing, headache, hypotension.

Disopyramide
Used in cardiac irregularities. Adverse effects include dry mouth, blurred vision, urinary hesitancy.

Distigmine
Anticholinesterase: see **Neostigmine.**

Disulfiram
Blocks alcohol metabolism at stage of acetaldehyde. Produces nausea, vomiting, severe headache, chest pain, dyspnoea, hypotension and collapse if taken before alcohol. Used in treatment of alcoholism. Other adverse effects include impotence, neuropathy and interference with anticoagulant activity of **Warfarin.**

Dithranol
Used topically in psoriasis and other chronic skin conditions. Adverse effects include severe irritation to the eyes and skin.

Domiphen
Topical disinfectant used in skin and mouth preparations.

Dopamine
Naturally occurring precursor of **Noradrenaline.** Used intravenously in treatment of shock where it is claimed to increase cardiac output without increasing peripheral resistance.

Dothiepin
Tricyclic antidepressant with actions, uses, etc. similar to **Imipramine.** Also has mild tranquillising action, which may be useful in aggitated depression. May cause extrapyramidal adverse effects.

Doxapram
C.N.S. stimulant used to stimulate respiration. Adverse effects include convulsions and cardiac irregularities.

Doxepin
Tricyclic antidepressant with actions, uses, etc. similar to **Imipramine.** Also has mild transquillising effect which may relieve anxiety associated with depression.

Doxorubicin
Cytotoxic antibiotic used in neoplastic disease. Adverse effects include bone marrow depression, cardiotoxicity, gastro-intestinal disturbances.

Doxycycline
Bacteriostatic antibiotic with actions, adverse effects and interactions similar to **Tetracycline.**

Doxylamine
Antihistamine similar to **Promethazine.**

Droperidol
Butyrophenone tranquilliser with actions and adverse effects similar to **Haloperidol.** Used in combination with analgesics such as **Phenoperidine** to maintain the patient in a state of neuroleptanalgesia—calm and indifferent whilst conscious and able to cooperate with the surgeon.

Drostanolone
Anabolic steroid given by intramuscular injection. Adverse effects as for **Testosterone.**

D-Thyroxine
Lowers blood cholesterol in hypercholesterolaemia but does not have the metabolic actions of **Thyroxine.** May cause angina in susceptible patients and may increase the activity of anticoagulant drugs, e.g. **Warfarin.**

Dydrogesterone
Actions similar to **Progesterone,** but does not inhibit ovulation and does not have contraceptive effect.

Dyflos
Organophosphorus long-acting anticholinesterase with actions, etc. similar to **Physostigmine.**

E

Ecothiophate
Anticholinesterase similar to **Dyflos.**

Edrophonium
Short-acting anticholinesterase with actions similar to **Physostigmine.** Used in diagnosis of myasthenia gravis.

Embramine
Antihistamine similar to **Promethazine.**

Emepronium
Parasympatholytic with actions, toxic effects, etc. similar to **Atropine.** Used to reduce tone in the urinary bladder when this is responsible for pain and urinary frequency.

Emetine
Anti-amoebic agent given by subcutaneous injection. Adverse effects include nausea, vomiting, hypotension, cardiac arrhythmias.

Ephedrine
Sympathomimetic amine with alpha and beta adrenoceptor effects. Bronchodilator used in bronchial asthma. Also as mydriatic and nasal decongestant. Adverse effects include tachycardia, anxiety, insomnia.

Epsom salts
See **Magnesium sulphate.**

Ergometrine
Ergot derivative that contracts the uterus. Used after delivery to prevent or reduce haemorrhage. Adverse effects as for **Ergotamine.**

Ergotamine
Ergot derivative with vasoconstricting and alpha adrenoceptor-blocking activity. Used in treatment of migraine by oral, intramuscular, sublingual, aerosol or suppository routes. Adverse effects include nausea, vomiting, headache, convulsions, cold extremities.

Erythromycin
Bactericidal antibiotic with spectrum of activity similar to **Benzylpenicillin,** plus some strains of H-influenzae and mycoplasmas. Adverse effects include diarrhoea. Erythromycin estolate can produce liver damage with jaundice.

Essential oils
Volatile oils taken orally for carminative effects in gastric discomfort. Induces a feeling of warmth with increased salivation. Large doses are irritant and may cause both gastro-intestinal symptoms and inflammation of the urinary tract.

Etafedrine
Similar to **Ephedrine.**

Etamiphylline
Bronchodilator with actions similar to **Aminophylline.**

Ethacrynic acid
Potent diuretic. Action and uses similar to **Frusemide.** Adverse effects similar to **Bendrofluazide.** May also cause transient deafness.

Ethambutol
Antituberculous drug. Well tolerated but high doses toxic to optic nerve, producing central or periaxial retrobulbar neuritis.

Ethamivan
Respiratory stimulant essentially similar to **Nikethamide.** May be used in respiratory depression of the newborn.

Ethamsylate
Haemostatic agent used to control surgical and menstrual blood loss.

Ethchlorvynol
Tertiary alcohol. Sedative/hypnotic. Rapid but short-lived general depressant action on C.N.S. Used in treatment of insomnia. Often leaves an 'after-taste'. May cause giddiness, weakness, depression and 'hangover'. Stimulates its own metabolism, danger of drug interactions. Coma in overdosage with severe respiratory depression. No antidote. Treated by supportive measures.

Ethiazide
Diuretic with actions similar to **Chlorothiazide.**

Ethinyloestradiol
Synthetic female sex hormone with similar actions and adverse effects to **Dienoestrol.** Combined with progestational drug in some oral contraceptives.

Ethionamide
Antituberculous agent. High incidence of adverse effects, mainly on gastro-intestinal tract.

Ethisterone
Similar actions and adverse effects to **Progesterone.**

Ethoglucid
Cytotoxic agent. Has been used in the treatment of various malignant conditions. Adverse effects include nausea, baldness, oedema.

Ethoheptazine
Analgesic for mild to moderate pain. Adverse effects include nausea and drowsiness.

Ethomoxane
Alpha adrenoceptor-blocking drug similar to **Phentolamine**.

Ethopropazine
Parasympatholytic used in treatment of Parkinsonism. Less effective than **Benzhexol** and causes more frequent side effects.

Ethosalmide
Analgesia with similar actions and adverse effects to **Salicylamide**.

Ethosuximide
Anticonvulsant. Supresses epileptic discharges in the brain. Used in treatment of petit mal (absence seizures) but not for major epilepsy. May cause nausea and vomiting, drowsiness or excitation, photophobia and Parkinson-like symptoms. Coma with respiratory depression in overdosage. No antidote. Supportive treatment only.

Ethotoin
Anticonvulsant essentially similar to **Phenytoin** but less toxic and less effective.

Ethyl biscoumacetate
Anticoagulant drug with actions similar to **Warfarin**.

Ethylmorphine
Analgesic, cough suppressant relate to **Morphine**. Adverse effects similar to morphine.

Ethyl nicotinate
Topical vasodilator. See **Nicotinic acid**.

Ethyloestrenol
Anabolic steroid. Adverse effects as for **Testosterone**.

Ethynodiol
Similar actions and adverse effects to **Progesterone**. Combined with oestrogenic agent in some oral contraceptives.

Eucalyptus
Essential oil used internally to relieve catarrh and externally as rubefacient.

Eucatropine
Parasympatholytic mydriatic similar to **Homatropine**.

F

Factor VIII
Blood clotting factor that is deficient in haemophilia and Von Willebrand's disease. Used intravenously to stop episodes of uncontrollable bleeding.

Fazadinium
Muscle relaxant with actions, adverse effects and uses similar to **Tubocurarine.**

Fencamfamin
C.N.S. stimulant. Adverse effects include anxiety and restlessness.

Fenfluramine
Anti-obesity with central anorectic and peripheral metabolic effects. May produce diarrhoea, sedation and sleep disturbance. Contra-indicated in patients taking monoamine oxidase inhibitors.

Fennel
Essential oil used as carminative.

Fenoprofen
Anti-inflammatory/analgesic with similar actions and uses to **Indomethacin.**

Fenoterol
Actions and adverse effects similar to **salbutamol.**

Fentanyl (c)
Narcotic analgesic with actions and uses similar to **Morphine.** More potent analgesic and respiratory depressant but shorter action.

Ferric ammonium citrate
Actions and adverse effects similar to **Ferrous sulphate.**

Ferric hydroxide
Iron salt with actions similar to **Ferrous sulphate.**

Ferrous fumarate
Actions and adverse effects similar to **Ferrous sulphate.**

Ferrous gluconate
Actions and adverse effects similar to **Ferrous sulphate.**

Ferrous glycine sulphate
See **Ferrous sulphate.**

Ferrous succinate
Actions and adverse effects similar to **Ferrous sulphate.**

Ferrous sulphate
Used in iron-deficiency anaemia. Adverse effects include black faeces, abdominal pain, constipation, diarrhoea. Liquid formulations can stain teeth black.

Flavoxate
Antispasmodic used in bladder disorders. Adverse effects include headache and dry mouth.

Fluclorolone
Topical **Corticosteroid** used in psoriasis and eczema.

Flucloxacillin
Antibiotic. Similar properties to **Cloxacillin** but better absorbed.

Flucytosine
Antifungal agent active orally against systemic candida infections. Adverse effects include bone marrow depression.

Fludrocortisone
Potent salt-retaining **Corticosteroid** used in adrenal insufficiency. Adverse effects include oedema, hypertension, electrolyte imbalance.

Flufenamic acid
Anti-inflammatory/analgesic essentially similar to **Mefenamic acid.**

Flumethasone
Topical **Corticosteroid** used in psoriasis and eczema.

Fluocinolone
Topical **Corticosteroid** used in psoriasis and eczema.

Fluocinonide
Topical **Corticosteroid** used in psoriasis and eczema.

Fluocortolone
Topical **Corticosteroid** used in psoriasis and eczema.

Fluorouracil
Cytotoxic drug used in neoplastic disease. Adverse effects include nausea, vomiting, diarrhoea, stomatitis, alopecia, skin pigmentation, bone marrow depression.

Fluoxymesterone
Male sex hormone used for deficiency states and as anabolic agent. Adverse effects similar to **Testosterone.**

Flupenthixol
Tranquilliser with antidepressant and anxiolytic actions but little sedative effects. Used in depressive and anxiety states associated with inertia and apathy. Adverse effects include restlessness, insomnia, hypotension, extra-pyramidal disturbances. Not recommended for children or excitable patients or in advanced cardiac, renal or hepatic disease.

Fluphenazine
Phenothiazine tranquilliser similar to **Chlorpromazine** but longer acting. Used in treatment of psychoses, confusion and agitation. Oral treatment required only once a day. Available as a 'depot' intramuscular injection which is active for 10–28 days. Adverse effects similar to Chlorpromazine but more frequently causes extrapyramidal disturbances.

Fluprednylidene
Topical **Corticosteroid,** used in psoriasis and eczema.

Flurandrenolone
Topical **Corticosteroid** used in psoriasis and eczema.

Flurazepam
Benzodiazepine tranquilliser/hypnotic. Used in the treatment of insomnia. Essentially similar to **Nitrazepam.**

Fluspirilene
Tranquilliser used in schizophrenia. Adverse effects include involuntary movements and low blood pressure.

Folic acid
Used in folate-deficient megaloblastic anaemias of pregnancy, malnutrition and malabsorption states. May precipitate neuropathy in untreated **Hydroxocobalamin** deficiency.

Formaldehyde
As a solution used topically for treatment of warts.

Framycetin
Antibiotic derivative of **Neomycin** used topically for skin infections and by mouth for gastro-enteritis and bowel sterilisation.

Frangula
Mild purgative with actions etc similar to *Cascara.*

Frusemide
Potent diuretic that causes greater reduction in sodium reabsorption by the kidney than occurs with the thiazide diuretics (see **Bendrofluazide**). Rapid onset of action when given orally or intravenously. Used in emergency treatment of fluid overload, especially pulmonary oedema and in cases resistant to thiazides. Adverse effects similar to Bendrofluazide.

Furazolidone
Poorly absorbed antibacterial drug used in bacterial diarrhoea and gastro-enteritis. Adverse effects include nausea, vomiting, rashes, haemolysis in predisposed patients and flushing with alcohol.

Fusafungine
Antibiotic administered by aerosol for infections of upper respiratory tract.

Fusidic acid
Steroid antibiotic used for infections by **Penicillin**-resistant staphylococci. Adverse effects include nausea and vomiting.

G

Gallamine
Skeletal muscle relaxant used during surgical procedures under general anaesthesia.

Gamma-benzene hexachloride
Applied topically for treatment of lice, scabies and other infestations. Adverse effects include convulsions if ingested.

Gefarnate
Recently introduced for treatment of peptic ulcer. May cause skin rashes.

Gelatin
Protein used as a nutrient in the preparation of some oral medicines and suppositories, and in a sponge-like form as a haemostatic.

Gentamicin
Bactericidal aminoglycoside antibiotic with spectrum similar to **Neomycin,** but specially active against *Pseudomonas aeruginosa.* Adverse effects include ototoxicity and nephrotoxicity, particularly in renal failure. Potentiates neuromuscular blockade.

Gestronol
Hormone with similar actions to **Progesterone.**

Glauber's salts
See **Sodium sulphate.**

Glibenclamide
Oral antidiabetic drug with actions and uses similar to **Chlorpropamide.**

Glibornuride
Oral antidiabetic drug with actions and uses similar to **Chlorpropamide.**

Glipizide
Oral antidiabetic drug with actions and uses similar to **Chlorpropamide.**

Glucagon
Polypeptide hormone produced by alpha cells of pancreas. Causes increase in blood sugar, release of several other hormones and increases force of cardiac contraction. Used in tests of carbohydrate metabolism and in treatment of heart failure. May cause nausea and vomiting but cardiac arrhythmias are said not to occur.

Glutaraldehyde
As a solution used to treat warts.

Gluten
Constituent of wheat starch responsible for bowel disorders in gluten-sensitive individuals. These conditions respond to treatment with a gluten-free diet.

Glutethimide
Hypnotic/sedative closely related to the barbiturates. Generalised depressant action on C.N.S. and some anticholinergic effects. Used in treatment of insomnia. May cause 'hangover', blurring of vision and gastric irritation. As with the barbiturates, tolerance and addiction may occur. Induces its own metabolism with danger of drug interaction. Coma with respiratory depression in overdosage. No antidote. Treatment is supportive.

Glycerine suppositories
Local lubricant purgative.

Glycerophosphates
Used widely in 'tonic' preparations as a source of phosphorus.

Glyceryl guaiacolate
See **Guaiphenesin.**

Glyceryl trinitrate
Vasodilator used sublingually in angina pectoris. Adverse effects include headache, dizziness, flushing.

Glycine
Amino acid used with antacids in gastric hyperacidity, and with aspirin to reduce its gastric irritation.

Glycol salicylate
Rubefacient. Essentially similar to **Salicylic acid.**

Glycopyrronium
Anticholinergic similar to **Atropine,** used in peptic ulcer and gastric hyper-acidity.

Glymidine
Oral antidiabetic with actions and uses similar to **Chlorpropamide.**

Gold salts
Anti-inflammatory agent, apparently specific for rheumatoid arthritis. Mechanism of action unknown. Given as a course of intramuscular injections. Toxic reactions are common including stomatitis, dermatitis, nausea, vomiting and diarrhoea. May cause hepatitis, nephritis and bone marrow depression. Not given if evidence of pre-existing liver or kidney disease.

Gonadorelin
Pituitary hormone that stimulates synthesis and release of sex hormones. Used in differential diagnosis of delayed puberty and hypogonadism.

Gonadothrophin
Pituitary hormone that stimulates gonadal activity. Used in infertility and delayed puberty.

Gramicidin
Antibiotic used by local application to skin, wounds, burns, and nose and mouth infections. Toxic if ingested or injected.

Grindelia
Used as expectorant and antispasmodic. Adverse effects include kidney irritation in large doses.

Griseofulvin
Antibiotic active against fungal infections of skin and nails when taken orally. May require higher doses in patients on anticonvulsant drugs.

Guaiphenesin
Used to reduce sputum viscosity.

Guanethidine
Adrenergic neurone-blocking drug. Used in hypertension. Eye drops used in glaucoma and hyperthyroid eye signs. Adverse effects include postural hypotension, nasal stuffiness, diarrhoea, fluid retention and impotence. Action antagonised by tricyclic antidepressants and sympathomimetics, e.g. when used as nasal decongestants in 'cold cures'.

Guanoclor
Antihypertensive adrenergic neurone-blocking drug with actions similar to **Guanethidine.**

Guanoxan
Antihypertensive adrenergic neurone-blocking drug with actions similar to **Guanethidine.**

Guar gum
Binding agent in tablets, thickening agent in foods.

H

Halcinonide
Topical **Corticosteroid** used in psoriasis and eczema.

Haloperidol
Butyrophenone tranquilliser. Used in treatment of psychosis where it has similar effects to **Chlorpromazine** but more potent. Has anti-emetic action but lacks anticholinergic and alpha adrenolytic effects. May cause Parkinsonism, drowsiness, depression, hypotension, sweating, skin reactions and jaundice. In overdosage effects and treatment similar to Chlorpromazine.

Halopyramine
Antihistamine similar to **Promethazine.**

Halothane
Inhalational anaesthetic.

Halquinol
Anti-amoebic/antibacterial/antifungal. Used in amoebic and bacterial diarrhoea. Adverse effects include allergic reactions.

Heparin
Anticoagulant produced in mast cells. Given parenterally. May produce allergic reactions and, on prolonged use, osteoporosis. Heparin-induced haemorrhage controlled by **Protamine sulphate.**

Heparinoid
See **Heparin.**

Heptabarbitone
Barbiturate hypnotic essentially like **Amylobarbitone.**

Heptaminol
Used as cardiac stimulant in bronchitis.

Heroin (c)
See **Diamorphine.**

Hexachlorophane
Topical antiseptic used in soaps, creams, lotions and dusting powders. Adverse effects include allergy and light sensitivity, and C.N.S. effects if absorbed or ingested.

Hexamethonium
Ganglion-blocking drug used parenterally in hypertension. Adverse effects include postural hypotension, dry mouth, paralysis of accommodation, retention of urine, constipation, impotence.

Hexamine
Antiseptic used topically and for urinary infections. For the latter use the urine must be rendered acid by giving also **Ammonium chloride** which liberates formaldehyde from the hexamine. May cause painful micturition, frequency and haematuria.

Hexamine mandelate
Compound of hexamine and mandelic acid, used as urinary antiseptic. Requires acid urine. Adverse effects include nausea and vomiting.

Hexetidine
Topical antibacterial/antifungal/antitrichomonas.

Hexobarbitone
Barbiturate hypnotic essentially like **Amylobarbitone.**

Hexylresorcinol
Antiworm. Also used as antiseptic agent for throat infections.

Histamine
Mediator of many body functions including gastric secretion, inflammatory and allergic responses. Produces skin vasodilatation. Used in test of gastric acid production. Adverse effects include headache, hypotension, bronchospasm, diarrhoea.

Homatropine
Parasympatholytic with actions, toxic effects, etc. similar to **Atropine.** Used as a mydriatic because when compared with atropine its action is more rapid, less prolonged and more easily reversed by **Physostigmine.**

Hyaluronidase
Enzyme that assists dispersal and absorption of subcutaneous and intramuscular injections. Hastens resorption of blood and fluid in body cavities. Adverse effects include allergic reactions.

Hydrallazine
Vasodilator antihypertensive drug. Adverse effects include tachycardia, headache, marrow depression, acute rheumatoid syndrome, S.L.E. syndrome.

Hydrargaphen
Mercurial antibacterial/antifungal used topically for skin and ear infections. Adverse effects include allergic reactions.

Hydrochlorothiazide
Thiazide diuretic similar to **Bendrofluazide**.

Hydrocortisone
Naturally occurring adrenocorticosteroid hormone with similar actions, etc. to **Cortisone**.

Hydroflumethiazide
Thiazide diuretic similar to **Bendrofluazide**.

Hydrotalcite
Antacid used in peptic ulcer and gastric hyperacidity.

Hydroxocobalamin
Vitamin B_{12} for the treatment of pernicious anaemia or specific deficiency states. Parenteral.

Hydroxyamphetamine
Sympathomimetic amine: see **Amphetamine**.

Hydroxychloroquine
Antimalarial agent: see **Chloroquine**.

Hydroxyprogesterone
Actions and uses similar to **Progesterone**.

Hydroxyquinoline
Topical antibacterial/antifungal deodorant.

Hydroxyurea
Cytotoxic agent for oral administration.

Hydroxyzine
C.N.S. depressant. Used to relieve tension and anxiety in emotional disturbances but less effective than **Chlorpromazine** and similar tranquillisers in the psychoses. May cause excessive drowsiness, headache, dry mouth, itching and convulsions. Coma in overdosage. No antidote; supportive treatment only.

Hyoscine butylbromide
Parasympatholytic with peripheral actions similar to **Atropine** but of shorter duration. Used as an antispasmodic similar to **Propantheline** but effective only by injection.

Hyoscine hydrobromide
Parasympatholytic with central and peripheral actions similar to **Atropine** except that it produces central depression and hypnosis rather than stimulation and that it tends to slow the heart. Used for pre-operative medication where the hypnotic effect makes it preferable to atropine and as an antiemetic for travel sickness. Adverse effects, etc. otherwise as for Atropine.

Hyoscine methobromide
Similar to **Hyoscine hydrobromide.**

Hyoscyamine
See **Atropine.**

Hypromellose
Similar to **Methylcellulose.**

I

Ibuprofen
Anti-inflammatory/analgesic. Used in rheumatoid arthritis. Gastro-intestinal side effects less common than with **Acetylsalicylic acid.** Headache and other C.N.S. symptoms have been described.

Ichthammol
Dermatological preparation with slight antibacterial effects. Used in creams and ointments for chronic skin conditions.

Idoxuridine
Antiviral agent used in local treatment of herpes infections.

Imipramine
Antidepressant, blocks neuronal re-uptake of **Noradrenaline,** dopamine and 5-hydroxytryptamine. Adverse effects include anticholinergic actions of dry mouth, blurred vision, precipitation of glaucoma, retention of urine, constipation; also produces cardiac arrhythmias, potentiates direct sympathomimetic pressor amines and antagonises action of **Guanethidine, Bethanidine, Debrisoquine** and **Clonidine.** Coma, convulsions and cardiac arrhythmias in overdosage. Treatment supportive.

Indomethacin
Anti-inflammatory/analgesic, used in treatment of inflammatory joint disease. Adverse effects include headache, vertigo, depression, confusion and gastrointestinal symptoms including perforation and haemorrhage.

Indoramin
Alpha adrenoceptor blocking drug, used in hypertension, peripheral vascular disease and prophylaxis of migraine. Produces sedation and nasal stuffiness.

Inositol nicotinate
Dilates peripheral blood vessels. Used for chilblains and other conditions where peripheral blood circulation is thought to be poor. Large doses may cause fall in blood pressure and slowing of heart.

Insulin
Hormone. Causes a fall in blood-sugar level and increased storage of glycogen in the liver. Used parenterally to treat diabetes. Adverse effects include hypoglycaemia, subcutaneous fat atrophy.

Ipecacuanha
Plant extract used in small doses as an expectorant in cough mixtures. Emetic effect if larger doses (syrup of ipecacuanha) are used in children as emergency treatment of ingested poisons.

Iprindole
Antidepressant with actions, uses and adverse effects similar to **Imipramine.**

Iproniazid
Monoamine oxidase inhibitor antidepressant. Actions, uses and adverse effects as for **Phenelzine.**

Iron dextran injection
Parenteral formulation for iron-deficiency anaemia. Adverse effects include pain on injection, skin staining, vomiting, headache, dizziness. Anaphylactic reactions may accompany intravenous infusion particularly.

Iron sorbitol injection
Intramuscular formulation for iron deficiency anaemia. Adverse effects as for **Iron dextran injection.**

Isoaminile citrate
Cough suppressant used on its own or in cough linctus. No analgesic or sedative effects. Does not depress respiration. May cause dizziness, nausea, constipation or diarrhoea.

Isocarboxazid
Monoamine oxidase inhibitor antidepressant. Actions, uses and adverse effects as for **Phenelzine.**

Isoetharine
Bronchodilator. Actions, uses and adverse effects as for **Salbutamol.**

Isometheptene
Sympathomimetic agent with actions and adverse effects similar to **Adrenaline.** Used in symptomatic treatment of migraine where it is said to constrict the dilated blood vessels that cause the throbbing headache.

Isoniazid
Synthetic antituberculous agent. About 60% of Caucasians are slow inactivators by acetylation, genetically determined. Adverse effects include peripheral neuropathy, pellagra, mental disturbances and convulsions, which may be reduced by administration of **pyridoxine.**

Isoprenaline
Beta adrenoceptor agonist used in bronchial asthma by inhalation or orally. Adverse effects include tachycardia, arrhythmias, tremor.

Isopropamide
Anticholinergic with actions and adverse effects similar to **Atropine.** Used in treatment of gastro-intestinal colic, peptic ulceration and as a decongestant in symptomatic relief of the common cold.

Isosorbide diniytrate (sorbide nitrate)
Dilates blood vessels. Similar actions and adverse effects to **Glyceryl trinitrate** but longer action. Used for symptomatic and prophylactic treatment of angina.

Isoxsuprine
Beta adrenoceptor agonist, produces uterine relaxation. Used in premature labour. Adverse effects include tachycardia, arrhythmias, tremor.

Ispaghula
Purgative. Increases faecal bulk. Mechanism of action similar to that of **Methylcellulose.**

K

Kanamycin
Bactericidal aminoglycoside antibiotic with actions, spectrum similar to **Neomycin,** but less ototoxic. Used in gram-negative septicaemia, with monitoring of blood levels, particularly in renal failure. Potentiates neuro-muscular blockade.

Kaolin
Adsorbent. Used externally as a dusting powder and by mouth as treatment for diarrhoea where it increases faecal bulk and slows passage through the gut. Once thought to have specific adsorbent effect for poisonous substances but it is now known that the adsorbent effect is a general one.

Ketoprofen
Anti-inflammatory/analgesic with actions, uses and adverse effects similar to **Ibuprofen.**

L

Labetalol
Antihypertensive. Has both alpha and beta adrenoceptor-blocking actions. Uses and adverse effects similar to **Propranolol.** The alpha-blocking action may be the main mechanism that lowers the blood pressure. Postural hypotension may occur.

Lactic acid
Used intravenously as dilute solution in treatment of acidosis. Acts less rapidly than **Sodium bicarbonate.** Also used topically as strong solution in treatment of warts.

Lactulose
Purgative. A synthetic disaccharide (galactose plus fructose) that is not absorbed but broken down by gut bacteria to non-absorbable anions that increase the faecal mass by osmotic effects. Effective but expensive. Has been recommended for use in liver failure to reduce absorption of ammonia from the gut.

Laevodopa
Amino acid. Converted in body to **Dopamine,** a neurotransmitter substance that is deficient in Parkinson's disease. Controls rigidity and improves movements but less effect on tremor than anticholinergic drugs, e.g. **Benzhexol.** May cause gastro-intestinal symptoms, hypotension, involuntary movements and psychiatric disturbances. Side effects may be reduced by combination with inhibitors of dopamine breakdown, e.g. **Carbidopa.** Contra-indicated/caution in cardiovascular disease and psychiatric disturbance. Effects diminished by phenothiazines (e.g. **Chlorpromazine),** **Methyldopa, Reserpine, Pyridoxine.**

Laevulose
Carbohydrate. Used intravenously as a source of calories when oral feeding is not possible. In renal failure it is better tolerated than dextrose. Accelerates metabolism of ethyl alcohol and may be used to treat alcohol poisoning. May cause facial flushing, abdominal pain and localised thrombophlebitis.

Lanatoside C
Foxglove derivative with actions, etc. similar to **Digoxin.**

L-Asparaginase (Colaspase)
Cytotoxic enzyme derived from bacterial culture; used in neoplastic disease. Adverse effects include nausea, vomiting, pyrexia, neurotoxicity, hypersensitivity reactions, bone marrow depression.

Lecithins
Phospholipids found in both animal and vegetable foods. Used as emulsifying and stabilising agents in skin preparations.

Leucinocaine
See **Panthesine.**

Levallorphan
Narcotic antagonist similar to **Nalorphine.** Less likely to cause severe withdrawal symptoms in 'addicts'.

Levodopa
See **Laevodopa.**

Levorphanol (c)
Narcotic analgesic similar to **Morphine** but more reliable when given by mouth. Useful in the management of severe chronic pain in terminal disease.

Lignocaine
Local anaesthetic/antidysrhythmic. Stabilises nerve cell membranes to prevent impulse conduction. Used topically or by injection for local anaesthesia in minor operations. Intravenous injection or infusion used to treat abnormal heart rhythms. Excessive doses block also motor impulses and normal cardiac conduction. May cause hypotension, C.N.S. depression and convulsions. Metabolised by liver and therefore used with caution in liver disease. Short action prevents use as oral antidysrhythmic.

Lincomycin
Antibiotic active against **Penicillin-**resistant staphylococci and bacteroides. Adverse effects include rashes, diarrhoea, pseudomembranous colitis.

Liothyronine
Thyroid hormone, probably the active hormone to which **Thyroxine** is converted. Given by mouth or injection it has effects similar to **Thyroxine** but more rapid and short-lived. Used when rapid effect is needed, e.g. in myxoedema coma. Used with care if there is evidence of cardiovascular disease as it may precipitate cardiac failure.

Liquid paraffin
Purgative. Lubricates faecal material in colon and rectum. Used when straining is undesirable or defaecation painful, e.g. after operations for haemorrhoids. Reduces absorption of fat-soluble **Vitamins A** and **D,** and in chronic use can cause paraffinomas in mesenteric lymph glands. May leak from anal sphincter.

Liquorice
Dried plant root with expectorant and mild anti-inflammatory properties.
Used as a flavouring/expectorant in cough mixtures. Deglycyrrhizinised
extract is used in treatment of peptic ulceration. Large doses may cause salt
and water retention leading to hypertension and/or cardiac failure.

Lithium salts
Usually given as carbonate or citrate, provides lithium ions which substitute
for sodium in excitable tissues and reduces brain catecholamine levels. Used
in treatment of mania. Caution in cardiac or renal disease. Needs careful
control of plasma levels. Adverse effects include tremor, vomiting, diarrhoea,
ataxia, blurred vision, thirst, polyuria leading to confusion, fits and coma in
gross overdosage. Lithium excretion may be enhanced by forced alkaline
diuresis, peritoneal dialysis, or haemodialysis.

Liver extracts
Extracts of liver prepared for oral use were used for treatment of pernicious
anaemia. Unpalatable and irregularly absorbed. Now replaced by the pure
vitamin B_{12}, **Hydroxocobalamin.**

Loperamide
Antidiarrhoeal with actions, uses and adverse effects similar to **Diphenoxylate.**

Lorazepam
Benzodiazepine anxiolytic similar to **Diazepam.**

Lymecycline
Bacteriostatic antibiotic with actions, adverse effects and interactions similar
to **Tetracycline.**

Lynoestrenol
Sex hormone (progestogen) with actions, uses and adverse effects similar to
Norethisterone.

Lypressin
Hormone extract from posterior pituitary gland of pigs. Actions, uses and
adverse effects similar to **Vasopressin.**

M

Mafenide
Sulphonamide antibacterial with actions similar to **Sulphadimidine**. Used topically on skin and as eye drops.

Magnesium antacids
Range of magnesium salts used alone or complexed with other compounds. Neutralise gastric acid in treatment of peptic ulceration. Large doses have laxative effect which may be reduced by combination with **Aluminium antacids**. Very little absorbed but danger of toxic magnesium blood levels in renal failure. May reduce absorption of other drugs, e.g. **Tetracyclines**.

Magnesium carbonate
Non-systemic antacid with similar actions, uses and adverse effects to **Magnesium hydroxide**. Releases carbon dioxide in stomach and may cause belching.

Magnesium hydroxide
Non-systemic antacid (only 10% absorbed). Used in treatment of peptic ulceration. Neutralises gastric acid and acts longer than **Sodium bicarbonate**. May have laxative effect, which can be prevented by simultaneous use of **Aluminium antacids.**

Magnesium oxide
Non-systemic antacid. Converted to **Magnesium hydroxide** in the stomach and has similar actions and adverse effects.

Magnesium sulphate (Epsom salts)
Saline purgative. Absorbed only slowly from gut. Magnesium and sulphate ions attract or retain water by osmosis and thus increase bulk of intestinal contents. Effective in 3–6 hours. Produces semi-fluid or watery stools, therefore useful as single treatment but not for repeated dosage. Danger of systemic toxicity from magnesium in patients with reduced renal function.

Magnesium trisilicate
Non-systemic antacid used in treatment of peptic ulceration. Neutralisation of acid is slow in onset but relatively prolonged due to adsorbent properties of silicic acid formed in the stomach. Has a laxative effect in larger doses. Danger of magnesium toxicity in patients with renal failure.

3

Malathion
Organophosphorus insecticide. Acts by inhibition of cholinesterase and may therefore produce toxic effects due to accumulation of excess **Acetylcholine**. One of the least toxic of this group of insecticides, low concentrations of malathion may be used on human skin for infestation (e.g. lice) without systemic effects. Toxic effects may be treated by antidotes **Atropine** and **Pralidoxime.**

Malic acid
An acid found in apples and pears. Formerly used in tooth-cleaning tablets. Used as part of an astringent skin treatment.

Manganese sulphate
Occasionally used as a haematinic. Said to increase the effect of **Ferrous sulphate** in treatment of iron-deficiency anaemia.

Mannitol
Osmotic diuretic. Opposes reabsorption of water which normally accompanies sodium reabsorption from kidney tubule. Used when there is danger of renal failure (e.g. shock, cardiovascular surgery) and in fluid overload refractory of other diuretics. May cause cardiac failure owing to increased circulating blood volume.

Mannomustine
Cytotoxic drug with actions, uses and adverse effects similar to **Mustine hydrochloride.**

Maprotiline
Antidepressant with actions, uses and adverse effects similar to **Imipramine.**

Mazindol
Anorectic, indole derivative with central stimulant properties. Produces tachycardia and rise in blood pressure.

Mebendazole
Used in treatment of roundworm. Actions and adverse effects as for **Thiabendazole.**

Mebeverine
Antispasmodic with direct action on colonic smooth muscle but no systemic anticholinergic effects. Used for relief of abdominal pain and cramps, e.g. due to irritable colon or non-specific diarrhoea.

Mebhydrolin
Antihistamine with uses and adverse effects similar to **Promethazine** but shorter duration of action and less likely to cause sedation.

Mecamylamine
Ganglion-blocking drug. Use and adverse effects as for **Hexamethonium.**

Meclofenoxate
C.N.S. stimulant claimed to be of benefit in treatment of mental confusion and other intellectual impairments. Must not be given to excitable patients.

Meclozine
Antihistamine with similar actions, uses and adverse effects to **Promethazine.**

Medazepam
Benzodiazepine anxiolytic similar to **Diazepam** but with less anticonvulsant activity. Used in the treatment of anxiety.

Medigoxin
Actions, uses and adverse effects similar to **Digoxin.**

Medroxyprogesterone
Sex hormone with actions, uses and adverse effects similar to **Progesterone.**

Mefenamic acid
Anti-inflammatory/analgesic. Mode of action uncertain. Used in treatment of arthritis. May cause severe diarrhoea. Other adverse effects include gastro-intestinal bleeding, exacerbation of asthma, haemolytic anaemia, bone marrow depression. May enhance action of oral anticoagulants (e.g. **Warfarin).**

Mefruside
Diuretic essentially similar to **Bendrofluazide.**

Melphalan
Cytotoxic drug used in myelomatosis. Actions and adverse effects similar to **Chlorambucil.**

Menadiol
Orally active form of **Vitamin K.**

Menapthone
See **Vitamin K.**

Menthol
Aromatic oil used as inhalation, orally as pastilles for relief of respiratory symptoms, or topically on skin where it causes dilatation of blood vessels producing a sense of coldness and analgesia.

Mepenzolate
Anticholinergic with actions and adverse effects similar to **Atropine.** Marked effect upon spasm of colon. Used to relieve pain, distension and diarrhoea associated with gastro-intestinal disorders.

Meperidine (c)
U.S.A.: see **Pethidine**.

Mephenesin
Muscle relaxant with actions on spinal cord and centrally. In low doses causes some relaxation of spastic muscles without reduction in power. Used for symptomatic relief of spastic conditions. May cause sedation, anorexia, nausea and vomiting. Intravenous injections have caused haemolysis, hypotension and kidney damage.

Mephenytoin
U.S.A.: see **Methoin**.

Mepivacaine
Local anaesthetic with actions, adverse effects and uses similar to **Lignocaine** but not used to treat abnormal heart rhythms.

Meprobamate
Minor tranquilliser (anxiolytic) with selective action on hypothalamus and spinal cord. Used in treatment of neuroses, alcoholism and functional disorders such as tension headache. May cause gastro-intestinal disorders, headache, dizziness with hypotension, lowered tolerance to alcohol and withdrawal symptoms. Induces hepatic drug metabolism with danger of drug interactions. Coma with respiratory depression in overdosage. No antidote. Forced alkaline diuresis and haemodialysis may be effective.

Mepyramine
Antihistamine with similar uses and adverse effects to **Promethazine** but has shorter duration of action and is less likely to cause sedation.

Mequitazine
Antihistamine with actions and adverse effects similar to **Promethazine**. Used in treatment of allergic conditions.

Mercaptopurine
Cytotoxic drug, inhibiting nucleoprotein synthesis, used in neoplastic disease, particularly leukaemia. Adverse effects include bone marrow depression.

Mercuramide
Mercury-containing diuretic with actions, uses and adverse effects similar to **Mersalyl**.

Mersalyl
Organic mercurial diuretic. Depresses active reabsorption of sodium and chloride by kidney tubules. Used in treatment of fluid retention. Long-acting; must be administered by intramuscular injection. Danger of excessive loss of sodium and chloride. May cause gastro-intestinal disturbance, skin rashes and, after prolonged use, kidney damage.

Mesterolone
Sex hormone with actions, uses and adverse effects similar to **Testosterone**.

Mestranol
Sex hormone with actions, uses and adverse effects similar to **Oestradiol**.

Metaraminol
Sympathomimetic agent with alpha and beta effects similar to **Adrenaline**. Alpha effects predominate and thus it has been used to raise blood pressure in hypotension after myocardial infarction. May cause headache, dizziness, nausea, vomiting and tremor. Overall, seldom has any lasting, worth-while effect.

Metformin
Oral antidiabetic agent with similar actions and uses to **Phenformin**.

Methacholine
Parasympathomimetic drug with muscarinic actions of **Acetylcholine**.

Methacycline
Antibacterial with actions, uses and adverse effects similar to **Tetracycline**.

Methadone (c)
Synthetic narcotic analgesic. Actions similar to **Morphine** but less sedation, euphoria and respiratory depression. Used in control of withdrawal symptoms from narcotic addiction and for relief of chronic pain in terminal disease. **Naloxone** is true antagonist.

Methallenoestril
Sex hormone with actions, uses and adverse effects similar to **Oestradiol**.

Methandienone
Sex hormone with actions and adverse effects similar to **Testosterone** but has greater anabolic effects for similar androgenic effects. Used together with adequate diets for the anabolic effects in conditions characterised by muscle and bone wasting.

Methapyrilene
Antihistamine with actions, uses and side effects similar to **Promethazine** but of shorter duration.

Methaqualone (c)
Hypnotic/sedative. General depressant action on C.N.S. Used in treatment of insomnia. Frequently has 'hangover' effect. May also cause localised loss of sensation with numbness and tingling as well as skin rashes and gastro-intestinal disturbance. Liable to abuse for so-called 'aphrodisiac' qualities and euphoriant effects. In overdose causes respiratory depression with increased muscle tone and increased reflexes. No antidote. Treatment is supportive.

Methdilazine
Antihistamine with actions, uses and adverse effects similar to **Promethazine** but more potent and said to be less sedative.

Methicillin
Antibiotic. Similar properties to **Cloxacillin,** but only active parenterally.

Methionine
Amino-acid, essential constituent of diet. May be used as an antidote in severe poisoning due to **paracetamol** where it is thought to prevent liver damage by reducing the concentration of a toxic metabolite of paracetamol. Given orally it causes few side effects, principally nausea. Must not be given more than 10 hours after the overdose as it may then exacerbate liver damage.

Methisazone
Antiviral agent used in prophylaxis against smallpox. Adverse effects include anorexia, nausea, vomiting, diarrhoea.

Methixene
Parasympatholytic used in treatment of Parkinsonism. Actions, etc. similar to **Benzhexol** but greater effect in reducing tremor.

Methocarbamol
Muscle relaxant with actions, uses and adverse effects similar to **Mephenesin.**

Methohexitone
Ultra-short-acting barbiturate hypnotic. Used for induction of anaesthesia. Actions and adverse effects similar to **Thiopentone sodium.**

Methoin (Mephenytoin—U.S.A.)
Anticonvulsant essentially similar to **Phenytoin** but probably more toxic, both in normal dosage and in overdosage.

Methoserpidine
Antihypertensive. Actions, uses and adverse effects similar to **Reserpine.**

Methotrexate
Cytotoxic drug, antagonising folic acid, used in neoplastic disease, particularly leukaemia. Adverse effects include alopecia, stomatitis, liver toxicity, folate-deficient anaemia, bone marrow depression.

Methotrimeprazine
Phenothiazine tranquilliser essentially similar to **Chlorpromazine.**

Methoxamine
Sympathomimetic amine, predominantly alpha adrenoceptor agonist. Produces vasoconstriction with rise of blood pressure. Used in hypotensive states. Toxicity includes hypertension, cerebral haemorrhage, pulmonary oedema.

Methoxyflurane
Potent volatile anaesthetic. Administered by inhalation. Actions and adverse effects similar to **Halothane.**

Methoxyphenamine
Sympathomimetic amine with predominantly beta effects. Used for prevention of asthma. Adverse effects include tachycardia, tremor, nausea, insomnia.

Methsuxamide
Anticonvulsant essentially similar to **Ethosuximide.**

Methylcellulose
Purgative. Indigestible plant residue. Absorbs water, increases faecal bulk and thus promotes bowel movements. Slow action ($\frac{1}{2}$–3 days). No important systemic effects.

Methylclothiazide
Diuretic essentially similar to **Bendrofluazide.**

Methylcysteine
Mucolytic with actions, uses and adverse effects similar to **Acetylcysteine.** Used orally as well as by aerosol inhalation.

Methyldopa
Antihypertensive. Reduces sympathetic tone by central and peripheral mechanisms. Adverse effects include sedation, depression, nasal stuffiness, fluid retention, impotence, haemolytic anaemia.

Methylephedrine
Sympathomimetic agent with actions, uses and adverse effects similar to **Ephedrine.**

Methylpentynol
Tertiary alcohol. Hypnotic/sedative. Rapid but short-lived generalised depression of the C.N.S. Used in treatment of insomnia and anxiety. May cause repeated belching. In overdosage symptoms resemble inebriation with alcohol. No antidote. Supportive treatment is adequate.

Methylphenidate (c)
Sympathomimetic agent with central stimulant effects similar to **Amphetamine.** Has little effect on appetite. Used only for its stimulant effects, e.g. in narcolepsy. May cause headache, gastro-intestinal symptoms and tremor.

Methylphenobarbitone
Anticonvulsant/sedative essentially similar to **Phenobarbitone.**

Methylprednisolone
Corticosteroid with actions, uses and adverse effects similar to **Prednisone.**

Methyprylone
Hypnotic sedative. Related to **Glutethimide** and essentially similar to that drug and **Amylobarbitone.**

Methyl salicylate
Rubefacient used for relief of musculo-skeletal pain. Has similar actions and adverse effect to **Acetylsalicylic acid** but not used systemically.

Methysergide
Serotonin antagonist used in preventive treatment for severe migraine. May cause nausea, abdominal cramp, dizziness and psychiatric disturbance. Prolonged use may cause retroperitoneal fibrosis resulting in impairment of renal function.

Methyltestosterone
Sex hormone with actions, uses and adverse effects similar to **Testosterone.**

Metoclopramide
Anti-emetic with central effects on brain and peripheral effects on gastro-intestinal tract where it stimulates motility to improve gastric emptying and intestinal transit. May cause drowsiness and muscle spasms. Used to treat nausea and vomiting from most causes and as an adjunct to X-ray examination of the gut.

Metolazone
Diuretic essentially similar to **Bendrofluazide.**

Metoprolol
Beta adrenoceptor-blocking drug, with limited cardioselectivity. Uses, side effects, etc. as for **Propranolol.**

Metronidazole
Antimicrobial. Effective against trichomonas, Vincent's organisms, giardi-osis and amoebiasis. Adverse effects include nausea, metallic taste in mouth, hypersensitivity reactions, **Disulfiram**-like reaction with alcohol.

Metyrapone
Inhibits enzyme responsible for synthesis of adrenocorticosteroids. Used in tests of pituitary gland function. May cause gastro-intestinal disturbance and dizziness.

Mexenone
Absorbs ultraviolet light and protects skin from sunburn.

Mexiletine
Local anaesthetic and antidysrhythmic agent similar to **Lignocaine** but also effective when given by mouth. May cause nausea, vomiting, drowsiness, tremors, convulsions, hypotension and bradycardia.

Mianserin
Antidepressant with uses similar to **Imipramine,** but without its peripheral autonomic adverse effects.

Miconazole
Antifungal agent used topically for skin infections.

Minocycline
Antibacterial with actions, uses and adverse effects similar to **Tetracycline.**

Mistletoe
Plant extract with vasodilator action. Has been recommended for treatment of hypertension and included in mixture for relief of bronchospasm.

Mitobronitol
Cytotoxic drug with actions, uses and adverse effects similar to **Busulphan.**

Monosulfiram
Parasiticide used topically for treatment of fleas, lice, ticks and mites. May cause skin rashes. Related to **Disulfiram** and may have similar effects if ingested.

Morazone
Analgesic/anti-inflammatory agent. Used only in a compound preparation.

Morphine (c)
Poppy derivative. Centrally acting narcotic analgesic used for relief of severe pain. Other potentially useful effects include euphoria and cough suppression. Also has adverse effects, respiratory depression, nausea, vomiting, constipation, hypotension, and physical dependence and ('addiction') abuse. Coma with danger of death in overdosage. **Naloxone** is a specific antagonist.

Mustine hydrochloride (Nitrogen mustard)
Cytotoxic drug used in neoplastic disease. Adverse effects include nausea, vomiting, bone marrow depression.

Myristylbenzalkonium
Antiseptic used in throat lozenges for treatment of minor throat infections.

N

Naftazone
A systemic haemostat recommended for control of haemorrhagic conditions.

Naftidrofuryl
Peripheral vasodilator said to improve cellular metabolism and to relax muscle cells in blood vessel walls. Recommended for treatment of reduced peripheral blood circulation and dementia caused by reduced blood flow. May cause headache, insomnia and gastro-intestinal disturbance.

Nalidixic acid
Urinary antiseptic to which resistance readily occurs. May cause gastro-intestinal symptoms and allergic reactions. May exacerbate epilepsy and respiratory depression.

Nalorphine
Narcotic antagonist. Reverses effects of **Morphine** and other narcotic analgesics but less specific than **Naloxone** as has some analgesic activity of its own. Used for reversal of narcotic effects but not when due to **Pentazocine**. May cause hallucinations and thought disturbances. In 'addicts' causes severe withdrawal symptoms.

Naloxone
Narcotic antagonist. A true antidote to **Morphine** and other narcotic analgesics. Has no analgesic activity in its own right. Used for reversal of narcotic effects from any narcotic drug especially respiratory depression. Danger of severe withdrawal symptoms if given to those physically dependent ('addicted') to narcotics.

Nandrolone
Sex hormone with actions and adverse effects similar to **Testosterone** but anabolic effects greater than androgenic effects. Used in treatment of debilitating illness and carcinoma of the breast. Injection only, not active by mouth.

Naphazoline
Sympathomimetic agent with marked alpha-adrenergic activity. Used topically on nasal mucosa where its vasoconstrictor effect leads to reduced secretion and mucosal swelling, e.g. in allergic rhinitis. Prolonged use may lead to rebound nasal congestion and secretion. Not used systemically but oral overdosage would cause depression of nervous system and coma.

Naproxen
Anti-inflammatory/analgesic with actions, uses and adverse effects similar to **Ibuprofen.**

Natamycin
Antifungal used topically for fungal infections of the vagina and skin. Not used orally because poorly absorbed and causes gastric irritation. May cause skin rashes.

Neomycin
Bactericidal aminoglycoside antibiotic with spectrum similar to **Streptomycin,** but more active against staphylococci and proteus. Not used systemically because of risk of ototoxicity and nephrotoxicity. Used topically and orally for bowel sterilisation and in liver failure. May cause malabsorption if given for long period. Potentiates neuromuscular blockade.

Neostigmine (Prostigmin)
Anticholinesterase with actions similar to **Physostigmine.**

Niacinamide
Vitamin. See **Nicotinic acid.**

Nialamide
Antidepressant/monoamine oxidase inhibitor with actions, uses and adverse similar to **Phenelzine.**

Niclosamide
Used in treatment of tapeworms. Adverse effects are uncommon as it is not absorbed from the intestinal tract.

Nicotinamide
Vitamin: see **Nicotinic acid.**

Nicotinic acid (Vitamin B_7)
Vitamin. Deficiency causes pellagra, with dermatitis, diarrhoea and dementia. Has direct relaxant effect on muscle in peripheral blood vessels and has been used to treat reduced peripheral circulation. Large doses may cause flushing and gastro-intestinal symptoms.

Nicotinyl tartrate
Peripheral vasodilator acting directly on muscle in blood vessels. Used for treatment of impaired peripheral blood circulation including chilblains and Raynaud's syndrome. May cause flushing, tachycardia, shivering, gastro-intestinal symptoms and fall in blood pressure.

Nicoumalone
Anticoagulant with actions, interactions and adverse effects similar to **Warfarin.**

Nifenazone
Analgesic/anti-inflammatory/antipyretic. Used orally or as suppositories in treatment of rheumatoid conditions. May cause nausea, vomiting and blood dyscrasias.

Nifuratel
Antiprotozoal/antifungal used topically and systemically for infections of the vagina. May cause nausea and gastro-intestinal discomfort.

Nikethamide
Respiratory stimulant acting directly on brain respiratory centres. Seldom useful except in respiratory depression due to severe chronic bronchitis. Not used in treatment of respiratory depression due to drug overdosage. May cause sweating, nausea, vomiting, convulsions and depression of the nervous system.

Nimorazole
Antiprotozoal used to treat certain gastro-intestinal and vaginal infections, e.g. giardiasis, trichomoniasis. Contra-indicated in neurological disease. Causes nausea if alcohol taken during treatment.

Niridazole (Ambilhar)
Anti-infective, used in schistosomiasis, guineaworms. Adverse effects include brown urine, psychoses, convulsions.

Nitrazepam
Benzodiazepine tranquilliser/hypnotic. Depressant action on C.N.S. Used in treatment of insomnia. Relatively 'safe'. No true addiction. May cause dizziness, unsteadiness and slurred speech. In overdosage respiratory depression much less severe than with barbiturates. No antidote; supportive treatment only.

Nitrofurantoin
Urinary antiseptic, producing yellow fluorescence in urine. Adverse effects include hypersensitivity reactions, nausea, vomiting, neuropathy; haemolytic anaemia in G.6 P.D.-deficient subjects.

Nitrofurazone
Antibacterial used topically for infection of the outer ear and skin wounds. Allergic reactions may occur.

Nitrogen mustard
See **Mustine hydrochloride.**

Nonoxynol
Non-ionic surfactant. Used as spermicidal cream.

Nonylic acid
Rubefacient used topically for musculo-skeletal pain.

Noradrenaline
Sympathomimetic amine, predominantly alpha adrenoceptor agonist. Produces general vasoconstriction with rise of blood pressure. Used in hypotensive states. Toxicity includes hypertension, cerebral haemorrhage, pulmonary oedema.

Norethandrolone
Sex hormone with actions similar to **Testosterone** but anabolic effects greater than androgenic effects. Active by mouth or by injection. Used to treat debilitating conditions and carcinoma of the breast. Used with caution if liver function is impaired.

Norethisterone
Sex hormone with actions and adverse effects of **Progesterone**. Used for oral contraception and treatment of uterine bleeding.

Norethynodrel
Sex hormone with actions and adverse effects similar to **Progesterone**. Used for oral contraception and treatment of uterine bleeding.

Norgestrel
Sex hormone with actions and adverse effects similar to **Progesterone**. Used for oral contraception.

Nortriptyline
Antidepressant with actions, uses and adverse effects similar to **Imipramine**.

Noscapine
Cough suppressant with actions and adverse effects similar to **Pholcodine**.

Novobiocin
Antibiotic used for infections by **Penicillin**-resistant staphylococci. Adverse effects include hypersensitivity reactions with urticarial rashes, and kernicterus in the neonate.

Noxytiolin
Anti-infective with antibacterial and antifungal activity. Used topically for prevention or treatment of infections in bladder or other body cavities, e.g. after bladder operations.

Nystatin
Antibiotic used in treatment of candida infections of skin and mucous membranes, particularly mouth, alimentary tract and vagina.

O

Oestradiol

Naturally occurring sex hormone (oestrogen). Controls development and function of female sex organs, working in conjunction with **Progesterone**. Could be used for menstrual disorders, oestrogen deficiency, oral contraception and suppression of certain neoplastic disease but mainly superseded by related compounds. May cause withdrawal bleeding, breast development in the male, salt and water retention, nausea and vomiting, stimulation of tumours and venous thrombosis. Avoid use in patients with known risks of these effects.

Oestriol

Sex hormone (oestrogen) similar to **Oestradiol** but more active by mouth. Used mainly for menopausal disorders.

Oestrogen

Sex hormone: see **Oestriol** and **Oestrone**.

Oestrone

Sex hormone (oestrogen) similar to **Oestradiol**. Used mainly for menopausal disorders.

Oleandomycin

Antibiotic with similar actions and adverse effects to **Erythromycin**.

Opipramol

Tricyclic antidepressant with actions and uses similar to **Imipramine**. Mild tranquillising effect may be useful in anxiety associated with depression.

Opium tincture (c)

Mixture of poppy alkaloids. See **Morphine**.

Orciprenaline

Beta adrenoceptor agonist used in bronchial asthma. Adverse effects include tachycardia, arrhythmias, tremor.

Orphenadrine

Parasympatholytic/antihistamine. Used as antispasmodic in treatment of Parkinsonism. Actions, adverse effects, etc. as for **Benzhexol**.

Ouabain
Plant derivative with effects on the heart similar to those of **Digoxin** but only reliably active when given by injection when the onset of action is more rapid than for digoxin.

Oxazepam
Benzodiazepine anxiolytic similar to **Diazepam.**

Ox bile extract
Recommended for treatment of biliary deficiency. Probably of little use except that it increases bowel activity and helps to relieve constipation.

Oxedrine
Sympathomimetic with mainly alpha-adrenergic effects. Adverse effects similar to **Noradrenaline** but weaker and longer acting. Recommended for treatment of hypotension.

Oxethazaine
Surface-active, local anaesthetic. Added to some antacid mixtures with intention of adding an analgesic effect. Actions and adverse effects similar to **Lignocaine.**

Oxolinic acid
Urinary antiseptic for treatment of urine infections. May cause gastro-intestinal symptoms and C.N.S. stimulation. Contra-indicated in epilepsy.

Oxpentifylline
Vasodilator. Used in treatment of peripheral vascular disorders where it is thought to reduce blood viscosity and thus increase flow. May cause nausea, dizziness, flushing and hypotension.

Oxprenolol
Beta adrenoceptor-blocking drug with partial agonist activity (intrinsic sympathomimetic activity). Uses, side effects, etc. as for **Propranolol.**

Oxycodone (c)
Narcotic analgesic with actions, uses and adverse effects similar to **Morphine.**

Oxymetazoline
Sympathomimetic with marked alpha-adrenergic effects. Used topically on nasal mucosa as treatment for nasal congestion. Actions and adverse effects similar to **Naphazoline.**

Oxymetholone
Sex hormone with similar actions, uses and adverse effects to **Methandienone.**

Oxypentifylline
Vasodilator used in treatment of peripheral vascular disease, e.g. intermittent claudication, Raynaud's syndrome. May reduce blood pressure, lower blood glucose or cause gastro-intestinal symptoms. Caution needed if given with antihypertensives or insulin as it may increase their effects.

Oxypertine
Tranquilliser used in treatment of schizophrenia, other psychoses and anxiety neuroses. May cause drowsiness (high doses) or hyperactivity (low doses). Gastro-intestinal disturbances, hypotension and extrapyramidal disturbances less frequent than with the phenothiazines. Coma with respiratory depression in overdosage. No antidote; supportive treatment only.

Oxyphenbutazone
Anti-inflammatory/analgesic essentially similar to **Phenylbutazone.**

Oxyphencyclamine
Anticholinergic with actions and adverse effects similar to **Atropine.** Relatively long action. Used to reduce gastric acid output and to suppress gastro-intestinal colic.

Oxyphenisatin
Purgative that acts by direct stimulation of colonic muscle. Has been shown to cause liver damage and is now no longer available in the U.K.

Oxyphenonium
Anticholinergic with actions and adverse effects similar to **Atropine.** Used systemically to suppress gastric acid and intestinal spasm. Eye drops used to dilate the pupils.

Oxyquinoline
Has antibacterial, antifungal, deodorant and keratolytic properties. Used topically on skin or in vagina for minor infections and acne. Sensitivity rashes may occur.

Oxytetracycline
Bacteriostatic antibiotic with actions, adverse effects and interactions similar to **Tetracycline.**

Oxytocin
Hormone from the posterior pituitary gland. Causes contraction of uterus. Used for induction of labour. May cause fluid retention and uterine rupture with danger to foetus. Contra-indicated in toxaemia of pregnancy, placental abnormalities or foetal distress.

P

Pancreatic enzymes
Extracts of animal pancreas. Used to aid digestion of starch, fats and protein when there is pancreatic deficiency. May cause sore mouth and sensitivity reactions with sneezing, watery eyes or skin rashes.

Pancreatin
See **Pancreatic enzymes.**

Pancuronium
Muscle relaxant with actions and uses similar to **Tubocurarine.**

Panthesine (Leucinocaine)
Local anaesthetic with actions similar to lignocaine. Used as an injection in combination with protoveratrines where its hypotensive effects are said to help in the treatment of severe hypertension. See **Veratrum.**

Pantothenic acid
Considered a vitamin but no proven deficiency disease in man. No accepted therapeutic role but included in some vitamin mixtures.

Papaveretum (c)
Mixture of poppy derivatives; 50% is **Morphine** to which it is similar in all respects.

Papaverine
Muscle relaxant with action on involuntary muscle. Used in bronchodilator aerosol mixtures and as relaxant in mixtures for gastro-intestinal spasm. Low toxicity except intravenously when it may cause cardiac arrhythmias.

para-Aminobenzoic acid
Nutrient. Essential metabolites for certain bacteria. Sometimes included in vitamin mixtures but there is no evidence of a deficiency disease in man. Large doses may cause nausea, skin rashes and hypoglycaemia.

para-Aminosalicylic acid (P.A.S.)
Synthetic antituberculous agent, usually given as the sodium salt. Adverse effects include hypersensitivity reactions, nausea, vomiting, goitre and hypothyroidism, hepatitis.

Paracetamol (Acetaminophen—U.S.A.)
Analgesic/antipyretic. Inhibits synthesis of prostaglandins in the brain but, unlike **Aspirin** does not have this action in the periphery and therefore has no anti-inflammatory effect. Relieves mild pain but not inflammation. Overdosage may cause potentially fatal liver damage. Cysteamine and methionine may be of value as antidotes.

Paradichlorobenzene
Insecticide included in some ear drops. May irritate the skin. Inhalation or ingestion may cause drowsiness.

Paraldehyde
Anticonvulsant, hypnotic, sedative. Used in status epilepticus or disturbed patients. Administered by intramuscular injection. Dissolves plastic syringes: glass must be used. Irritant at injection site. Exhaled in breath producing unpleasant odour.

Paramethadione
Anticonvulsant essentially similar to **Trimethadione.**

Paramethasone
Synthetic corticosteroid with similar actions to **Cortisone** but has greater anti-inflammatory activity with less effect in salt and water retention.

Parathyroid hormone
Extract from parathyroid glands. Increases plasma calcium by releasing calcium from bone, reducing urine calcium excretion and increasing its absorption from the gastro-intestinal tract. Active only by injection. Used to treat low plasma calcium (tetany). May cause abnormally high plasma calcium with weakness, lethargy and coma.

Pargyline
Monoamine oxidase inhibitor. Reduces blood pressure by unknown mechanism. Antihypertensive. Shares adverse interactions of **Phenelzine,** and may produce fluid retention.

P.A.S.
See *Para*-**aminosalicylic acid.**

Pecilocin
Antifungal used topically in treatment of fungus infections of skin, nails and scalp.

Pectin
Emulsifying agent used in preparation of pharmaceutical and cosmetic products.

Pemoline
C.N.S. stimulant with action intermediate between **Caffeine** and **Amphetamine**. Used as treatment for lethargy, e.g. with epilepsy. May cause insomnia, anxiety and rapid heart rate.

Pempidine
Ganglion-blocking drug. Use and adverse effects as for **Hexamethonium**.

Penamecillin
Essentially similar to **Phenoxymethylpenicillin**.

Penicillamine
Chelating agent. Binds certain toxic metals, e.g. copper, and increases their excretion. Used in treatment of metal poisoning and rheumatoid arthritis (where its mechanism of action is uncertain). May cause headache, fever, loss of taste, gastro-intestinal symptoms, kidney damage and bone marrow depression.

Penicillins
Group of bactericidal antibiotics that act by inhibiting bacterial cell wall synthesis. Hypersensitivity cross-reactions occur to all.

Penicillin V
See **Phenoxymethylpenicillin**.

Pentaerythritol tetranitrate
Vasodilator with longer, but milder, action than **Glyceryl trinitrate** whose adverse effects it shares. Used to prevent angina attacks.

Pentagastrin
Synthetic gastro-intestinal hormone. Stimulates secretion of gastric acid, pepsin and pancreatic enzymes. Used by injection as test of gastric and pancreatic function. May cause nausea, flushing, dizziness and fall in blood pressure.

Pentamidine
Used in treatment of leishmaniasis. Adverse effects include hypotension, nausea, vomiting.

Pentazocine
Narcotic analgesic more potent than **Codeine** but less potent than **Morphine**. Dependence and addiction less common than with morphine. Some activity as a narcotic antagonist like **Naloxone**. May cause bad dreams, hallucinations and withdrawal symptoms. **Naloxone** but not **Nalorphine** nor **Levallorphan** may be used as antagonists.

Penthienate
Anticholinergic with actions and adverse effects similar to **Atropine**. Used to relieve gastro-intestinal spasm.

Pentifylline
Vasodilator with actions, uses and adverse effects similar to **Oxypentifylline.**

Pentobarbitone
Barbiturate hypnotic essentially like **Amylobarbitone.**

Pentolinium
Ganglion-blocking drug, used parenterally in hypertension. Adverse effects as for **Hexamethonium.**

Pepsin
Enzyme found in normal gastric juice. Controls breakdown of proteins. May be given to improve digestion when there is deficiency of pepsin secretion.

Perhexilene
Anti-anginal agent. Used for prevention of symptoms but not treatment of acute attack. May cause dizziness, nausea, vomiting, weakness, flushing and skin rashes.

Pericyazine
Tranquilliser with actions, uses and adverse effects similar to **Chlorpromazine.**

Perphenazine
Phenothiazine tranquilliser similar to **Chlorpromazine,** used in treatment of psychotic disorders, agitation and confusion. Not recommended for children.

Pethidine (Meperidine—U.S.A.) (c)
Synthetic narcotic analgesic essentially similar to **Morphine.** Used for relief of severe pain. May produce nausea, vomiting, dry mouth, euphoria, sedation and respiratory depression. Coma with danger of death on overdosage. **Naloxone** is a specific antagonist. Long-term use associated with tolerance, physical dependence and ('addiction') abuse.

Phenacemide
Anticonvulsant. Suppresses epileptic discharge in the brain. Used only when other anticonvulsants have proved ineffective, e.g. in temporal lobe epilepsy. Adverse effects are common, including aggressive behaviour, depression, drowsiness, gastro-intestinal disturbances, blood disorders. Coma and respiratory depression in overdosage. No antidote. Supportive treatment.

Phenacetin (Acetophenetidin—U.S.A.)
Analgesic/antipyretic but not anti-inflammatory. Converted by the liver to **Paracetamol** which is the main active form. Available only in combined

analgesic preparations. Does not cause gastric irritation but may cause haemolytic anaemia and methaemoglobinaemia. Has been associated with kidney damage (analgesic nephropathy) and is now seldom used in U.K.

Phenazocine (c)
Narcotic analgesic similar to **Morphine** but causes less sedation, vomiting and hypotension. More likely to depress respiration than morphine.

Phenazone
Analgesic/antipyretic. Local anaesthetic action when applied topically. Not much used but found in some analgesic mixtures and in ear drops. May cause skin rashes and bone marrow suppression. Overdose may result in nausea, coma and convulsions. The rate of metabolism (half-life) of single doses is sometimes used as a measure of drug metabolism by the liver.

Phenazopyridine
Analgesic recommended for relief of pain and irritation in the urinary tract, e.g. cystitis. Causes orange or red coloration of urine. May cause headache, dizziness and blood disorders. Contra-indicated in renal or liver failure.

Phenbenicillin
As for **Phenoxymethylpenicillin.**

Phenbutrazate (c)
Anorectic, sympathomimetic with actions, uses and adverse effects and abuses similar to **Amphetamine.**

Phenelzine
Antidepressant; inhibits monoamine oxidase thus increasing tissue concentrations of **Noradrenaline,** dopamine and 5-hydroxytryptamine. Adverse effects include hepatitis, interactions with tyramine-containing foods and indirect sympathomimetic amines producing hypertensive crises, and with narcotics to produce profound C.N.S. depression. Hypertensive crisis treated with alpha adrenoceptor blocker, e.g. **Phentolamine.**

Phenethicillin
Essentially similar to **Phenoxymethylpenicillin.**

Pheneturide
Anticonvulsant essentially similar to **Phenacemide.**

Phenformin
Oral antidiabetic. Increases use of glucose by peripheral tissues. Used alone or in combination with sulphonylureas (e.g. **Chlorpropamide**) or **Insulin.** Most useful in overweight subjects where it suppresses appetite. May cause nausea, vomiting and diarrhoea.

Phenindamine
Antihistamine with actions and uses similar to **Promethazine.** Unlike most antihistamines it causes stimulant side effects and may be used when sedation is a problem. May cause insomnia, convulsions, dry mouth and gastrointestinal disturbances.

Phenindione
Anticoagulant with actions and interactions similar to **Warfarin.** Adverse effects include allergic reactions, jaundice, steatorrhoea.

Pheniramine
Antihistamine with actions, uses and adverse effects similar to **Promethazine.**

Phenmetrazine (c)
Anorectic, sympathomimetic amine, widely abused, with actions and adverse effects of **Amphetamine.**

Phenobarbital
U.S.A.: see **Phenobarbitone.**

Phenobarbitone (Phenobarbital—U.S.A.)
Long-acting barbiturate anticonvulsant. Depresses epileptic electrical discharges in the brain. Used orally as preventive treatment in epileptics and occasionally by injection to control a severe fit. Danger of sedation and impairment of learning capacity. Induces its own metabolism by the liver. Given only with caution in liver disease. Danger of drug interaction. Coma with respiratory depression in overdose. No antidote. Treatment is supportive, sometimes plus forced alkaline diuresis or haemodialysis to promote excretion of the drug.

Phenol
Oily injection used in treatment of haemorrhoids.

Phenolphthalein
Purgative that acts by direct stimulation of colonic muscle. Action prolonged by absorption of drug from the gut and recirculation in bile. Produces pink urine and faeces (red if alkaline). May cause skin rashes in sensitive individuals.

Phenoperidine (c)
Narcotic analgesic with actions and adverse effects similar to **Morphine.** Used with **Droperidol** to produce neuroleptanalgesia—a state of consciousness but calmness and indifference allowing the patient to cooperate with the surgeon.

Phenoxybenzamine
Alkylating agent with alpha adrenoceptor blocking and antihistamine effects. Used in phaeochromocytoma. Side effects include sedation, nausea and vomiting.

Phenoxybenzylpenicillin (Phenbenicillin)
As for **Phenoxymethylpenicillin.**

Phenoxyethylpenicillin (Phenethicillin)
As for **Phenoxymethylpenicillin.**

Phenoxymethylpenicillin (Penicillin V)
Acid-resistant penicillin used orally. Shares actions and adverse effects of **Benzylpenicillin.**

Phenoxypropanol
Preservative/anti-infective for skin preparations.

Phenoxypropylpenicillin (Propicillin)
As for **Phenoxymethylpenicillin.**

Phensuximide
Anticonvulsant essentially similar to **Ethosuximide.**

Phentermine
Anorectic, sympathomimetic amine. Actions and adverse effects similar to **Diethylpropion.**

Phentolamine
Alpha adrenoceptor-blocking drug with partial agonist and smooth-muscle relaxant activity. Used in phaeochromocytoma. Side effects include tachycardia and nasal stuffiness.

Phenylbutazone
Anti-inflammatory/analgesic used in treatment of inflammatory joint disorders. Adverse effects include nausea, vomiting, skin rashes, peptic ulceration, sodium retention, hypotension. Occasionally causes bone marrow depression with thrombocytopaenia, agranulocytosis and aplastic anaemia. May enhance action of oral anticoagulants (e.g. **Warfarin**).

Phenylephrine
Sympathomimetic amine with actions, uses and toxicity of **Noradrenaline.** Also used as mydriatic and nasal decongestant.

Phenylethylbarbiturate
See **Phenobarbitone.**

Phenylmercuric nitrate
Has antibacterial, antifungal and spermicidal actions. Used as a preservative and as a chemical contraceptive.

Phenylpropanolamine
Sympathomimetic with actions, uses and adverse effects similar to **Ephedrine.**

Phenyltoloxamine
Antihistamine with actions, uses and adverse effects similar to **Promethazine.**

Phenytoin
Anticonvulsant. Suppresses epileptic discharge in the brain. Used orally to prevent convulsions and by injection to control convulsions or to suppress irregular heart rhythms. Long-term use may cause gum hypertrophy, acne, hirsutism, folate deficiency, anaemia, osteomalacia and liver enzyme induction with danger of drug interactions. In mild overdosage causes ataxia, dysarthria and nystagmus. Coma and respiratory depression in severe cases. No antidote. Supportive therapy only.

Pholcodine
Narcotic derivative related to **Codeine** but with little analgesic activity. Used only for cough suppression. May cause constipation. Similar to Codeine in overdosage.

Phosphoric acid
In dilute form acts as a stimulant to gastric secretion.

Phosphorylcolamine
Synthetic amino acid with high phosphorus content. Said to promote improved metabolism. Used as a 'tonic' in debilitated patients.

Phthalylsulpathiazole
Sulphonamide antibacterial with actions, etc. of **Sulphadimidine.** Poorly absorbed. Used mainly for gut infections and sterilisation.

Physostigmine (Eserine)
Anticholinesterase, allowing accumulation of acetylcholine. Actions those of **Acetylcholine.** Effects of overdose antagonised by **Atropine.**

Phytomenadione
See **Vitamin K.**

Pilocarpine
Parasympathomimetic drug with actions of **Acetylcholine.**

Pimozide
Tranquilliser with uses, adverse effects, etc. similar to **Chlorpromazine.**

Pindolol
Beta adrenoceptor-blocking drug with partial agonist activity (intrinsic sympathomimetic activity). Uses, side effects, etc. as for **Propranolol.**

Pipazethate
Cough suppressant with actions, uses and adverse effects similar to **Pholcodine.**

Pipenzolate
Anticholinergic with actions and adverse effects similar to **Atropine**. Used to reduce gastric acid secretion and intestinal spasm.

Piperazine
Used in treatment of threadworms and roundworms. Adverse effects include dizziness and ataxia.

Piperidolate
Anticholinergic with actions and adverse effects similar to **Atropine**. Relatively weak. Used to decrease gastro-intestinal motility and spasm.

Piritamide (c)
Narcotic analgesic with actions and adverse effects similar to **Morphine**. Injection only. Used for relief of post-operative pain.

Pituitary gland extract
Extract of animal pituitary tissue used for its antidiuretic activity. See **Vasopressin**.

Pizotifen
For prevention of migraine. Has antiserotonin and antihistamine properties. May cause drowsiness, weight gain, dizziness and nausea.

Podophyllum
Purgative. Pronounced irritant effect on the bowel or skin. Because of its violent effects has been replaced by milder drugs. Still used as a paint for warts where it prevents growth.

Poldine
Parasympatholytic with actions, etc. similar to **Atropine**. Used to reduce gastric acid secretion in treatment of peptic ulceration.

Poloxalene
Purgative. Lowers surface tension of intestinal fluids and softens faeces.

Polymyxin B
Antibiotic active against gram-negative bacteria. Not absorbed when taken by mouth but effective topically, e.g. within gut or on skin. May also be given by intramuscular injection. Rarely causes skin sensitivity but injections may be painful and associated with neurological symptoms.

Polynoxylin
Antiseptic with wide antibacterial and antifungal actions. Used topically for skin, throat and external ear infections.

Polysaccharide–iron complex
Haematinic with actions, uses and adverse effects similar to **Ferrous sulphate**.

Polythiazide
Thiazide diuretic similar to **Bendrofluazide.**

Posterior pituitary extract
Mixed hormonal extract with actions of **Oxytocin** and **Vasopressin** whose toxic effects it also shares. Used by injection or nasal absorption to treat diabetes insipidus.

Potassium aluminium sulphate
See **Alum.**

Potassium bicarbonate
Antacid. Has been used as gastric antacid as **Sodium bicarbonate,** but unsuitable for intravenous use.

Potassium chloride
Potassium supplement. Used when there is a danger of hypokalaemia, e.g. treatment with potassium-losing diuretics, fluid overload in liver failure. Oral potassium chloride itself causes nausea and gastric irritation. Usually administered as slow-release preparation or in an effervescent solution of bicarbonate and trimethylglycine. May be given by slow intravenous infusion. Danger of hyperkalaemia in renal failure. Treated by haemodialysis and ion-exchange resins.

Potassium citrate
Renders urine less acid. Used to reduce bladder inflammation. May produce adverse effects similar to **Potassium chloride.**

Potassium clorazepate
Anxiolytic, with actions, uses and adverse effects similar to **Diazepam.** Long acting and has sedative effects so is best given at night. Metabolised to desmethyldiazepam, an active metabolite of **Diazepam.**

Potassium gluconate
For treatment of potassium deficiency. Adverse effects similar to **Potassium chloride** but may be less irritant to the gastro-intestinal tract.

Potassium guaiacolsulphonate
Expectorant included in certain cough mixtures.

Potassium-*para*-aminobenzoate
Nutrient. See *para*-**Aminobenzoic acid.** Has been used to treat skin disorders where there is excessive fibrosis, e.g. scleroderma.

Potassium perchlorate
Treatment for overactive thyroid gland. Reduces formation of thyroid hormone by interfering with uptake of iodine into the gland. May cause nausea, vomiting, rashes, kidney damage and bone marrow suppression.

Povidone-iodine
Antiseptic. Liberates inorganic iodine slowly onto the skin or mucous membranes. Used pre-operatively and in treatment of wounds.

Practolol
Cardioselective beta adrenoceptor-blocking drug. Withdrawn because of adverse effects on eye, ear and peritoneum.

Pralidoxime (P2S)
Cholinesterase reactivator used in treatment of organophosphorus anti-cholinesterase poisoning.

Pramoxine
Surface-active, local anaesthetic with actions and adverse effects similar to **Lignocaine.** Used topically on skin or mucous membranes.

Prazosin
Vasodilator/antihypertensive. Adverse effects include tachycardia and headache. Excessive fall in blood pressure may occur early in treatment.

Prednisolone
Synthetic corticosteroid with actions, etc. as for **Prednisone.**

Prednisone
Synthetic corticosteroid with similar actions, etc. to **Cortisone** but has greater anti-inflammatory activity with less effect on salt and water retention.

Prenylamine
Vasodilator used to prevent angina attacks. May cause gastro-intestinal symptoms, flushing, skin rashes and hypotension. Contra-indicated in cardiac or liver failure.

Prilocaine
Local anaesthetic similar to **Lignocaine,** but less toxic. Used in dentistry.

Primaquine
Antimalarial agent. Adverse effects include nausea, methaemoglobinaemia, haemolytic anaemia.

Primidone
Anticonvulsant. Similar to barbiturates; partly metabolised by liver to phenobarbitone. Suppresses epileptic discharges in the brain. Used orally to prevent convulsions. Additive effect with **Phenobarbitone** to which it is otherwise essentially similar.

Probenecid
For prevention of gout. Increases urine excretion of uric acid and thus reduces its levels in the body. May cause nausea, vomiting and skin rashes. Should not be given with **Acetylsalicylic acid** whose actions it antagonises.

Pro

Procainamide
Antidysrhythmic/local anaesthetic similar to **Procaine** but longer acting and with less C.N.S. stimulation. Used to treat cardiac dysrhythmias but contra-indicated in heart block. May cause dose-related hypotension, mental depression and hallucinations. Hypersensitivity may cause arthritis and rash (systemic lupus erythematosis-like syndrome).

Procaine
Local anaesthetic. Stabilises nerve cell membranes to prevent impulse transmission. Used by injection for anaesthesia in minor operations. Poor activity if applied topically. Short action (due to rapid removal in the blood) may be prolonged by combination with a vasoconstrictor, e.g. **Adrenaline.** May cause C.N.S. stimulation with euphoria and convulsions. Metabolite of procaine interferes with antimicrobial activity of sulphonamides, e.g. **Sulphadimidine.** Preparations containing adrenaline are contra-indicated in heart disease, hyperthyroidism or treatment with tricyclic antidepressants, e.g. **Amitriptyline,** where it may cause cardiac dysrhythmias.

Procaine penicillin
Long-acting form of **Benzylpenicillin,** with similar actions and adverse effects.

Procarbazine
Cytotoxic drug used in neoplastic disease. Adverse effects include nausea, vomiting, diarrhoea, stomatitis, alopecia, neurotoxicity, bone marrow depression.

Prochlorperazine
Phenothiazine similar to **Chlorpromazine** but less sedative and more potent anti-emetic actions. Used mainly as an anti-emetic. More likely than chlorpromazine to cause extrapyramidal side effects. May be given orally, by injection or as suppository.

Procyclidine
Parasympatholytic used in treatment of Parkinsonism. Actions, etc. similar to **Benzhexol.**

Progesterone
Sex hormone acts on the uterus (in sequence with **Oestrogen**) to prepare the endometrium to receive the fertilised ovum. Has been used in treatment of uterine bleeding, for contraception, for breast and uterine tumours and for threatened abortion. Has to be injected and therefore largely replaced by newer progestational agents which are active by mouth. May cause acne, weight gain, enlargements of the breasts, gastro-intestinal symptoms and jaundice.

Proguanil
Antimalarial agent. Adverse effects include vomiting and renal irritation.

Prolintane
C.N.S. stimulant claimed to have effect intermediate between **Caffeine** and **Amphetamine**. Used to treat lethargy. May cause nausea, rapid heart rate and insomnia.

Promazine
Phenothiazine/antihistamine similar to **Chlorpromazine.**

Promethazine
Phenothiazine/antihistamine with actions similar to **Chlorpromazine.** Used as an anti-emetic and in treatment of allergic reactions but has little anti-psychotic effects. Marked sedative effects make it useful as a hypnotic in children and in pre-operative medication. Adverse effects and overdosage effects similar to **Chlorpromazine.**

Propamidine
Antiseptic with antibacterial and antifungal actions. Used topically for infections of skin and conjunctiva. Treatment should not be prolonged more than one week or tissue damage may occur.

Propanidid
Short-acting intravenous anaesthetic. Used for minor operations when quick recovery is needed. May cause nausea, vomiting, abnormal movements, spasm of larynx and respiratory depression.

Propantheline
Parasympatholytic with peripheral and toxic effects similar to **Atropine.** Used to reduce gastric acid secretion in peptic ulceration and as an anti-spasmodic for gastro-intestinal and urinary complaints. Contra-indications and overdosage effects as for atropine.

Propatylnitrate
Vasodilator with actions, uses and adverse effects similar to **Glyceryl trinitrate.**

Propicillin
Essentially similar to **Phenoxymethylpenicillin.**

Propranolol
Beta adrenoceptor antagonist used in angina, hypertension, arrhythmias, hyperthyroidism and anxiety. May cause bronchoconstriction, cardiac failure and cold extremities.

Propylene glycol
Solvent used in extract of some crude drugs and as a vehicle for some injec-tions and topical applications. May cause local irritation but less toxic than other glycols owing to rapid breakdown and excretion.

Propylhexedrine
Sympathomimetic used as inhalation for treatment of nasal congestion. Actions and adverse effects similar to **Naphazoline.**

Propyphenazone
Analgesic. Derivative of *Phenazone* with similar actions and adverse effects.

Protamine sulphate
Specific antidote to anticoagulant effect of **Heparin.** Derived from fish protein. Adverse effects include hypotension and dyspnoea.

Prothionamide
Antituberculous agent with actions and uses similar to **Ethionamide.** May be better tolerated.

Prothipendyl
Tranquilliser/anxiolytic/anti-emetic with similar uses, precautions and over-dosage effects to **Chlorpromazine.** Adverse effects include photosensitivity and convulsions.

Protoveratrines A and B
Plant-extract alkaloids. Antihypertensive but seldom used. See **Veratrum.**

Protriptyline
Antidepressant with similar action and adverse effects to **Imipramine** but has central stimulating effects. Used to treat depression associated with withdrawal and lack of energy. May aggravate anxiety and insomnia.

Proxymetacaine
Surface-active, local anaesthetic with actions and adverse effects similar to **Lignocaine.** Used in ophthalmology.

Proxyphylline
Bronchodilator with actions and uses similar to **Aminophylline.** Said to cause fewer gastro-intestinal symptoms.

Pseudoephedrine
Sympathomimetic with actions, uses and adverse effects similar to **Ephedrine.** Used mainly as decongestant. Said to have less effect on increasing blood pressure.

Psyllium
Purgative. Increases faecal bulk by same mechanism as **Methylcellulose.**

Pyrazinamide
Antituberculous drug. High incidence of adverse effects, particularly liver toxicity.

Pyridostigmine
Anticholinesterase with actions similar to **Physostigmine.**

Pyridoxine (Vitamin B$_6$)
Vitamin used in treatment of specific deficiency and other anaemias and in **Isoniazid**-induced neuropathy.

Pyrimethamine
Antimalarial agent. Adverse effects include skin rashes and folate-deficient anaemia.

Pyrrobutamine
Antihistamine with actions, uses and adverse effects similar to **Promethazine.**

Q

Quinalbarbitone
Barbiturate hypnotic usually prescribed in combined preparation with **Amylobarbitone**. No major differences from amylobarbitone.

Quinestradol
Sex hormone (oestrogen) similar to **Oestradiol**. Claimed to have greater effect on the vagina than on the uterus or breast. Used for post-menopausal vaginitis.

Quinestrol
Sex hormone (oestrogen) similar to **Oestradiol**. Used for suppression of lactation and menopausal disorders.

Quinethazone
Diuretic essentially similar to **Bendrofluazide**.

Quinidine
Antidysrhythmic agent with local anaesthetic activity. Depresses myocardial contractility and impulse conduction. Reduces cardiac output. Used to prevent recurrent dysrhythmias or to convert established dysrhythmias back to normal sinus rhythm. Dose-dependent effects include vertigo, tinnitus, deafness, blurred vision, confusion, gastro-intestinal symptoms, cardiac arrhythmias and cardiac arrest. Rashes and bruising are dose-independent. Contra-indicated when dysrhythmia is due to **Digoxin** or when there is heart block.

Quinidine phenylethylbarbiturate
General sedative/antidysrhythmic. Actions and adverse effects as for **Phenobarbitone** and **Quinidine**.

Quinine
Antimalarial agent. Adverse effects include vomiting, psychosis, visual and auditory disturbances, haemolytic anaemia and thrombocytopenia. Toxic doses may cause abortion.

R

Rauwolfia
Indian shrub. See **Reserpine** for main derivative.

Razoxane
Antimitotic used in treatment of certain bone and soft-tissue tumours together with radiotherapy. May cause gastro-intestinal disturbance, bone marrow suppression and hair loss.

Reserpine
Rauwolfia derivative. Reduces sympathetic tone by **Noradrenaline** depletion. Depletes brain noradrenaline, dopamine and 5-hydroxytryptamine. Used in hypertension and as anti-psychotic. Adverse effects include depression, Parkinsonism, nasal stuffiness, fluid retention, impotence.

Resorcinal
Dermatological treatment. Reduces itching and helps remove scaly skin. Used topically in treatment of acne and dandruff. Also as ear drops where used for antiseptic effects. If absorbed over long term, may cause suppression of thyroid gland. If ingested, is corrosive and may cause kidney damage, coma and convulsions.

Riboflavine (Vitamin B$_2$)
Vitamin. Deficiency leads to mucosal ulceration and angular stomatitis.

Rifampicin
Bactericidal antibiotic, used in tuberculosis. Adverse effects include liver toxicity and influenza-like symptoms. Induces liver enzymes, so reducing effectiveness of oral contraceptives and corticosteroids.

Rimiterol
Beta adrenoceptor agonist. Actions, uses, adverse effects as for **Salbutamol.**

Ritodrine
Beta adrenoceptor agonist. Actions, uses and adverse effects as for **Isoxsuprine.**

Rolitetracycline
Antimicrobial with actions and adverse effects similar to **Tetracycline** but not absorbed by mouth. Used by injection when an immediate high blood concentration is needed or when oral treatment is not possible.

Rubidomycin
See **Daunorubicin.**

4

S

Salbutamol
Bronchoselective beta adrenoceptor agonist used in bronchial asthma by inhalation, i.v. infusion or orally. Adverse effects include tachycardia, arrhythmias, tremors.

Salicylamide
Analgesic/antipyretic with actions and adverse effects similar to **Acetylsalicylic acid** but less effective and used only infrequently. In overdosage does not cause acidosis but depression of respiration and loss of consciousness.

Salicylic acid
Anti-inflammatory/analgesic; an active metabolite of **Acetylsalicylic acid** whose adverse effects it shares. Not used systemically as it causes marked gastric irritation. Topically on skin it acts as a keratolytic and has bacteriostatic and antifungal properties. Used to treat warts, skin ulcers, psoriasis and other skin conditions.

Salsalate
Anti-inflammatory/analgesic. After absorption is broken down to **Salicylic acid**. Uses and adverse effects similar to **Acetylsalicylic acid.**

Secbutobarbitone
Barbiturate hypnotic essentially similar to **Amylobarbitone.**

Selenium sulphide
Reduces formation and dandruff and other forms of eczema of the scalp. Used as a shampoo. Highly toxic if ingested causing anorexia, garlic breath, vomiting, anaemia and liver damage.

Senna
Plant extract purgative with actions, adverse effects, etc. as for **Cascara.**

Silver protein
Has mild antibacterial properties. Used in eye drops or nasal sprays for treatment of minor infections.

Silver sulphadiazine
Sulphonamide derivative with actions similar to **Sulphadimidine.** Used topically in treatment of burns to prevent infection.

Simethicone
Silicone/silica mixture used for water-repellent and antifoaming properties. Included in barrier creams and in oral treatments for flatulence.

Soap spirit
Soft soap in alcohol used in some dermatological preparations for its cleaning and descaling actions.

Sodium acid citrate
Anticoagulant. Now preferred to **Sodium citrate.**

Sodium acid phosphate
Saline purgative with actions and uses similar to **Sodium phosphate.**

Sodium alkyl sulphoacetate
Wetting agent/laxative. Used mainly as an enema for treatment of persistent constipation and pre-operative bowel evacuation.

Sodium antimonylgluconate (Triostam)
Used in schistosomiasis. Adverse effects include anorexia, nausea, vomiting, diarrhoea, muscle and joint pains, cardiotoxicity.

Sodium benzoate
Antiseptic/preservative included in mouth washes and for storage solutions for surgical instruments. Metabolised by liver to hippuric acid. Used as a test of liver function.

Sodium bicarbonate
Absorbable (systemic) antacid. Rapidly dissolves and neutralises acid in stomach. Produces quick relief of dyspepsia due to peptic ulceration but is not retained in stomach and therefore has short duration of action. Absorbed from small intestine, may cause systemic alkalosis. If used in large doses, with large doses of milk may cause renal damage—'milk–alkali syndrome'—. Danger of fluid retention in patients with cardiac failure or renal disease.

Sodium bromide
C.N.S. depressant used for sedative and calming effects. Now generally replaced by more effective, less toxic agents. Adverse effects include rashes, impaired mental function, anorexia, slurred speech. If treatment still continued this leads to ataxia, hallucinations and coma. Treatment of poisoning includes use of diuretics and haemodialysis.

Sodium calcium edetate
Chelating agent. Exchanges its calcium for other metal ions in the blood. Most effective exchange is for lead and it may be used by injection or by mouth for treatment of lead poisoning. May cause nausea, diarrhoea, abdominal cramps, pain and thrombophlebitis at site of injection. Renal damage and dermatitis have occurred with prolonged treatment. Used with caution if there is pre-existing renal disease.

Sodium cellulose phosphate
Non-absorbable powder taken by mouth in treatment of hypercalcaemia. Absorbs calcium ions in the intestine and prevents their absorption thus reducing the dietary intake of calcium.

Sodium chloride
Essential component of body fluids and tissues. Used intravenously to replace lost fluids when rapid treatment is needed or orally when replacement is less urgent, e.g. for sweat loss in tropics. Hyperosmolar solutions have been recommended as an emetic for first-aid treatment of poisoning but saline is a poor emetic and a good poison which may cause death due to hypernatraemia.

Sodium citrate
Mild purgative used in some enemas. Was used as an anticoagulant in blood for transfusion but now superseded by **Sodium acid citrate.**

Sodium cromoglycate
Preventive treatment for asthma, rhinitis and conjunctivitis due to allergy. Acts by blocking allergic mechanisms. Administered by inhalation of powder or topically in eye. May cause bronchial irritation and spasm and contact dermatitis.

Sodium edetate
Chelating agent used intravenously to reduce high blood calcium levels. Actions and adverse effects similar to **Sodium calcium edetate.** May cause excessive lowering of calcium levels.

Sodium iodide
Expectorant. Causes increased and more watery bronchial secretion. Included in some cough mixtures. Acts also as a source of iodine (essential for production of thyroid hormone). Added to table salt to prevent endemic goitre and may be used pre-operatively to prepare hyperactive goitre for removal. Should not be given in pulmonary tuberculosis where it may reactivate the disease.

Sodium iron edetate
Haematinic with actions, uses and adverse effects similar to **Ferrous sulphate**

Sodium morrhuate
Sclerosing agent used for injection treatment of varicose veins; causes obliteration of the dilated vessels. May cause allergic reactions. A test dose is recommended.

Sodium perborate
Mild disinfectant/deodorant used for mouth infections. Prolonged use may cause blistering and swelling in mouth.

Sodium phosphate
Saline purgative. Poorly absorbed from the gastro-intestinal tract. Retains water in the intestine and thus increases faecal mass. May be used orally or as suppository.

Sodium picosulphate
Saline purgative with actions and uses similar to **Magnesium sulphate.**

Sodium polystyrene sulphonate
Ion-exchange resin used in treatment of high plasma potassium levels where it exchanges sodium ions for potassium. May be used orally or rectally. Adverse effects include nausea, vomiting, constipation and sodium overload which may cause cardiac failure.

Sodium ricinoleate
Surface-active agent used in some toothpastes for its cleaning properties.

Sodium salicylate
Analgesic/anti-inflammatory/antipyretic with actions, uses and adverse effects similar to **Acetylsalicylic acid.** Usually taken in solution. Danger of sodium overload in patients with cardiac failure or renal failure.

Sodium sulphate (Glauber's salts)
Saline purgative. Actions and uses similar to **Magnesium sulphate.** Unpleasant taste. Danger of sodium retention with congestive heart failure in susceptible subjects.

Sodium tetradecyl sulphate
Injection used for treatment of varicose veins.

Sodium valproate
Anticonvulsant. May act by increasing brain levels of gamma-aminobutyric acid (GABA). Used in all forms of epilepsy. May cause gastro-intestinal symptoms and prolonged bleeding times with thrombocytopaenia. Before surgery, check for bleeding tendencies. May potentiate effects of antidepressant drugs whose dose should be reduced in combined treatment.

Sorbic acid
Preservative with antibacterial and antifungal properties.

Sorbide nitrate
See **Isosorbide dinitrate.**

Sorbitol
Carbohydrate poorly absorbed by mouth, but used as intravenous infusion it is a useful source of calories. May also be used as sweetening agent in diabetic foods or in dialysis fluids.

Sotalol
Beta adrenoceptor-blocking drug. Uses, side effects, etc. as for **Propranolol.**

Spectinomycin
Antimicrobial active against a wide range of bacteria. Offers no advantages over other antimicrobials except in treatment of gonorrhoea where a single injection may be adequate.

Spermicides
A number of different substances used in spermicidal contraceptives and administered topically into the vagina as jelly, creams, foaming tablet, pessary, aerosol or film. Appear to act by reducing surface tension on the sperm cell surface and allowing osmotic imbalance to destroy the cell. Relatively ineffective contraceptives, they should be used in conjunction with a barrier contraceptive, e.g. the cap, unless the couple concerned accept the risk of pregnancy.

Spiramycin
Antibiotic with similar actions and adverse effects to **Erythromycin.**

Spironolactone
Potassium-sparing diuretic. Acts by antagonism of the sodium-retaining hormone **Aldosterone** and thus prevents exchange of sodium for potassium in the kidney tubule. Diuretic action is weak. Used when aldosterone is an important cause of fluid overload, e.g. liver cirrhosis and nephrotic syndrome. Toxic effects include headache, nausea, vomiting and swelling of the breasts (especially in men). Danger of excessive potassium retention which, if severe is treated with haemodialysis and ion-exchange resins.

Squalane
Ingredient of skin ointments that increases skin permeability to drugs.

Squill
Expectorant. Has irritant action on gastric mucosa and produces reflex expectorant action. Used in cough mixtures for chronic bronchitis when sputum is scanty, but too irritant for use in acute bronchitis. May cause nausea, vomiting, diarrhoea and slowing of heart rate.

Stanolone
Sex hormone with actions, uses and adverse effects similar to **Testosterone.**

Stanozolol
Sex hormone with actions, uses and adverse effects similar to **Methandienone.** Long-term use may cause jaundice; used with caution in liver disease.

Sterculia
Purgative plant extract. Takes up moisture and increases faecal mass which promotes peristalsis.

Stibocaptate
Actions and adverse effects as for **Sodium antimonylgluconate.**

Stibogluconate sodium
Antimony derivative. Used in treatment of leishmaniasis. Adverse effects include nausea, vomiting, diarrhoea, muscle and joint pains, cardiotoxicity.

Stibophen
Used in schistosomiasis. Actions and adverse effects as for **Sodium antimonylgluconate.**

Stilboestrol
Sex hormone with actions, uses and adverse effects similar to **Oestradiol.**

Storax
Balsam obtained from trunk of *Liquidambar orientalis.* Has mild antiseptic action. Used topically to assist healing of skin, e.g. for bed sores and nappy rash.

Streptodornase
Enzyme derived from streptococcal bacteria. Breaks down proteins in exudates. Used together with **Streptokinase** to help remove clotted blood or fibrinous/purulent accumulations. Administered topically, intramuscularly or by instillation into body cavities, e.g. for haemothorax. May cause pain, fever, nausea, skin rashes and more severe allergic reactions. If haemorrhage occurs, the treatment is as for streptokinase.

Streptokinase
Plasminogen activator fibrinolytic agent derived from streptococcus. Given intravenously in thrombotic or embolic disease. May produce allergic reactions or haemorrhage which can be reversed by an antifibrinolysin such as **Aminocaproic acid** or **Tranexamic acid.**

Streptomycin
Bactericidal aminoglycoside antibiotic, active against tubercle bacillus, many gram-negative and some gram-positive organisms. Poorly absorbed orally. Administered intramuscularly. Excreted mainly by kidneys, so accumulates if renal function impaired. Adverse effects include hypersensitivity reactions (particularly contact dermatitis), ototoxicity, potentiation of neuromuscular blockade.

Strychnine
C.N.S. stimulant that is included in small doses in some 'tonics' with no justification. In fact is a potent poison causing convulsions, muscle spasms, coma and death. Treatment is by prevention of convulsions and assisted respiration.

Styramate
Centrally acting muscle relaxant with actions and uses similar to **Mephenesin.**
May cause drowsiness, dizziness and rashes.

Succinic acid
Said to promote absorption of iron from the intestine.

Succinylsulphathiazole
Sulphonamide antibacterial with actions, etc. of **Sulphadimidine.** Poorly
absorbed. Used mainly for gut infections and sterilisation of bowel prior to
surgery.

Sulfametopryrazine
Sulphonamide antibacterial with actions, etc. similar to **Sulphadimidine.** Long
acting. Side effects may be more serious than sulphadimidine.

Sulphacarbamide
Sulphonamide antibacterial with actions and adverse effects similar to
Sulphadimidine. Rapidly excreted in urine and therefore used for urinary-
tract infections. Crystalluria said not to occur.

Sulphacetamide
Sulphonamide antibacterial with actions similar to **Sulphadimidine,** but used
only as eye drops for eye infections.

Sulphadiazine
Sulphonamide antibacterial with actions, etc. similar to **Sulphadimidine.**

Sulphadimethoxine
Sulphonamide antibacterial with actions, etc. of **Sulphadimidine,** but only
once-daily administration required.

Sulphadimidine
Sulphonamide antibacterial which inhibits conversion of *para*-**Aminobenzoic
acid** to **Folic acid.** Broad spectrum of activity. Mainly used in urinary-
tract infections. Adverse effects include crystalluria, skin rashes, polyarteritis,
Stevens–Johnson syndrome. May produce kernicterus in newborn. Potentiates **Warfarin** by competitive displacement from plasma proteins.

Sulphafurazole
Sulphonamide antibacterial with actions, etc. similar to **Sulphadimidine.**

Sulphaguanidine
Sulphonamide antibacterial with actions, etc. of **Sulphadimidine.** Poorly
absorbed. Used mainly for gut infections and sterilisation of bowel prior to
surgery.

Sulphamerazine
Sulphonamide antibacterial with actions, etc. similar to **Sulphadimidine.**

Sulphamethizole
Sulphonamide antibacterial with actions, etc. similar to **Sulphadimidine.**

Sulphamethoxazole
Sulphonamide antibacterial with actions, etc. of **Sulphadimidine,** but somewhat longer action.

Sulphamethoxydiazine
Sulphonamide antibacterial with actions, etc. of **Sulphadimidine,** but only once-daily administration required.

Sulphamethoxypyridazine
Sulphonamide antibacterial with actions, etc. of **Sulphadimidine,** but only once-daily administration required.

Sulphanilamide
Sulphonamide antibacterial with actions and adverse effects similar to **Sulphadimidine** but more toxic. Now used only for topical infections, e.g. in eye or ear drops.

Sulphaphenazole
Sulphonamide antibacterial with actions, etc. of **Sulphadimidine,** but only once-daily administration required.

Sulphapyridine
Sulphonamide antibacterial with actions and adverse effects similar to **Sulphadimidine.** Toxic effects are common and its use is generally limited to treatment of dermatitis herpetiformis and other skin conditions.

Sulphasalazine
Compound of **Sulphapyridine** and **Salicylic acid,** used in ulcerative colitis. Adverse effects as for **Sulphadimidine.**

Sulphasomizole
Sulphonamide antibacterial with actions, etc. of **Sulphadimidine,** but somewhat longer action.

Sulphathiazole
Sulphonamide antibacterial with actions, etc. of **Sulphadimidine.**

Sulphinpyrazone
Prophylactic treatment for gout. Promotes renal excretion of urates by reducing reabsorption in renal tubules. Reduces blood uric acid levels and gradually depletes urate deposits in tissues. No value in treatment of acute gout. May cause nausea, vomiting and abdominal pain. May aggravate peptic ulcer and may precipitate acute gout. Long-term use may suppress bone marrow activity. Caution in renal disease and peptic ulcer. May interact to enhance actions of oral anticoagulants, oral hypoglycaemics and insulin.

Sulphormethoxine
Sulphonamide antibacterial with actions, etc. of **Sulphadimidine,** but only once-weekly administration required.

Sulphur
Used topically in skin lotions or ointments as an antiseptic.

Sulthiame
Anticonvulsant. Carbonic anhydrase inhibitor similar to **Acetazolamide.** Used in prevention of epilepsy, usually in addition to other drugs. May cause paraesthesiae of face and extremities, hyperventilation and gastric upsets. Inhibits **Phenytoin** metabolism and may cause phenytoin toxicity. In overdosage causes vomiting, headache, hyperventilation and vertigo but not coma. May cause crystalluria with renal damage which is treated by alkaline diuresis.

Suxamethonium
Muscle relaxant. Acts by depolarisation of muscle end plate rendering the tissue incapable of responding to the neurotransmitter. Action limited by destruction by pseudocholinesterase. Used as an adjunct to anaesthesia for surgery. Short acting but effects are prolonged in patients with reduced pseudocholinesterase levels. May cause bradycardia, cardiac arrhythmias, fever and bronchospasm. Prolonged respiratory paralysis is treated by assisted ventilation and *not* by anticholinesterases.

T

Tacrine
C.N.S. stimulant with anticholinesterase properties. Has been used as a respiratory stimulant, to reverse muscle relaxation due to **Tubocurarine,** or to prolong muscle relaxation due to **Suxamethonium.**

Talampicillin
Antimicrobial ester of ampicillin to which it is rapidly metabolised after absorption. Uses and adverse effects similar to **Ampicillin.** Said to maintain higher blood levels and to cause less diarrhoea.

Talc
Has lubricant and anti-irritant properties. Used topically on skin and as an aid to the manufacture of some tablets.

Tamoxifen
Anti-oestrogen. Competes with oestrogen for tissue receptor sites. Used as palliative treatment for breast cancer and in treatment of infertility due to failure of ovulation. May cause gastro-intestinal disturbance, fluid retention, hot flushes and vaginal bleeding.

Tannic acid
Astringent. Precipitates proteins and forms complexes with some heavy metals and alkaloids. May be used topically on skin for minor burns, abrasions or chilblains. Formerly used orally to reduce absorption of some poisons. May cause liver damage, nausea and vomiting.

Terbutaline
Beta adrenoceptor agonist. Actions, uses, adverse effects as for **Salbutamol.**

Terebene
Pleasant smelling oil used to mask unpleasant odours or tastes and as a vapour to relieve nasal decongestion. Large doses are irritant to the gastro-intestinal tract.

Testosterone
Male sex hormone. Controls development and maintenance of male sex hormones and secondary sex characteristics (androgenic effects). Also produces metabolic effects that lead to increased growth of bone, water

retention, increased production of red blood cells, and increased blood vessel formation in the skin (anabolic effects). Used in the male to speed sexual development, but of no value in treating sterility or impotence unless related to sexual underdevelopment. In the female used to treat some menstrual disorders, for suppression of lactation, and to reduce growth of breast tumours. Has also been used for anabolic effects in debilitated patients but now superseded by new drugs, e.g. **Methandienone.** Unwanted effects include excess fluid and water retention, stimulation of growth of prostate tumours, and virilisation in females.

Tetrachloroethylene
Used in treatment of hookworms. Adverse effects include nausea, vomiting, diarrhoea, vertigo.

Tetracosactrin
Synthetic polypeptide with actions, uses and adverse effects similar to **Corticotrophin.** Used intravenously as a test of adrenal function or by depot injection for treatment of inflammatory or degenerative disorders.

Tetracycline
Bacteriostatic antibiotic, active against many gram-positive and -negative organisms, some viruses and chlamydia. Adverse effects include diarrhoea and candida bowel superinfection, yellow discoloration of teeth and inhibition of bone growth in children, exacerbation of renal failure. Interacts in the bowel with compounds of iron, calcium and aluminium to produce insoluble chelates that are not absorbed.

Tetranicotinoylfructose
Peripheral vasodilator with actions similar to **Nicotinic acid.** Used to treat reduced peripheral circulation, e.g. Raynaud's syndrome. May cause flushing, rashes and heavy menstrual blood losses.

Thenyldiamine
Antihistamine with actions, uses and adverse effects similar to **Promethazine** but shorter duration of action.

Theobromine
Xanthine derivative. No useful C.N.S. stimulant effects. Has been used as a diuretic or to dilate coronary or peripheral arteries. Now superseded by more effective agents but still found in some mixtures.

Theophylline
Xanthine derivative. Smooth-muscle relaxant and diuretic with only minimal C.N.S. stimulant effects. Little used since aminophylline and other derivatives are more useful bronchial muscle relaxants. Is included with **Mersalyl** diuretic injection. Adverse effects as for **Aminophylline.**

Theophylline ethylenediamine
See **Aminophylline.**

Thiabendazole
Used in treatment of roundworms. Adverse effects include nausea, drowsiness, vertigo.

Thiambutosine
Antileprotic drug. Adverse effects include skin rashes and antithyroid action.

Thiamine
See **Aneurine.**

Thiethylperazine
Phenothiazine with actions similar to **Chlorpromazine** but little tranquillising effect. Used as an anti-emetic. Given orally, by injection or as suppository. Adverse effects, etc. as for chlorpromazine.

Thioacetazone
Antituberculous agent, used as a cheap alternative to *para*-**Aminosalicylic** acid. May cause gastro-intestinal symptoms, blurred vision, conjunctivitis and allergic reactions.

Thiocarlide
Antituberculous agent. Generally reserved for use in patients who are sensitive to other antituberculous drugs or who do not respond to such treatment. May cause rashes, joint pains and leucopaenia.

Thioguanine
Cytotoxic drug with actions, uses and adverse effects similar to **Mercaptopurine.**

Thiomersal
Mercurial disinfectant with antibacterial and antifungal actions. Used topically to prepare skin for operation. Also used in eye drops and for urethral irrigation. May cause hypersensitivity rashes.

Thiopentone sodium
Very short-acting barbiturate used intravenously for anaesthesia of short duration or induction of anaesthesia prior to use of other anaesthetics. Mode of action and adverse effects similar to **Amylobarbitone.**

Thiopropazate
Phenothiazine tranquilliser with actions, uses, etc. similar to **Chlorpromazine.**

Thioridazine
Phenothiazine tranquilliser similar to **Chlorpromazine.** Used in treatment of psychoses, confusion and agitation.

Thiotepa
Cytotoxic drug used in neoplastic disease. Adverse effects include bone marrow depression.

Thiothixene
Tranquilliser essentially similar to **Chlorpromazine.**

Thonzylamine
Antihistamine with actions, uses and adverse effects similar to **Promethazine** but shorter duration of action.

Threitol dimethane sulfonate
Cytotoxic drug used for treatment of ovarian cancer. May cause gastro-intestinal disturbance, bone marrow suppression and allergic rashes.

Thromboplastin
Blood-clotting factor, usually extracts of cattle brains used in a test that measures blood coagulability.

Thymoxamine
Alpha adrenoceptor-blocking drug used in peripheral vascular disease and glaucoma. Produces sedation and nasal stuffiness on intravenous administration.

Thyrotrophin
Pituitary hormone that stimulates production of thyroid hormones. Used in tests of thyroid function.

Thyroxine
Thyroid hormone. Has a stimulating action in general metabolism which is delayed in onset and prolonged (see **Liothyronine**). Used in treatment of thyroid deficiency. Doses in excess of requirements may cause thyrotoxic symptoms, e.g. rapid pulse, cardiac arrhythmias, diarrhoea, anxiety features, sweating, weight loss and muscular weakness. Caution if there is pre-existing heart disease.

Tigloidine
Chemically similar to **Atropine** but lacks most anticholinergic effects. Recommended for treatment of muscular rigidity and spasticity.

Timolol
Beta adrenoceptor antagonist with actions, uses and adverse effects similar to **Propranolol.**

Titanium dioxide
Dermatological treatment. Reduces itching and absorbs ultraviolet rays. Used topically to prevent sunburn and to treat some forms of eczema.

Tobramycin
Antimicrobial. Actions and adverse effects similar to **Gentamicin**.

Tocopheryl
See **Vitamin E**.

Tofenacin
Antidepressant with actions and uses similar to **Imipramine**. Recommended for elderly patients but may have marked anticholinergic effects with danger of urinary retention and glaucoma.

Tolazamide
Antidiabetic with actions, uses and adverse effects similar to **Chlorpropamide**.

Tolazoline
Alpha adrenoceptor-blocking drug with partial agonist activity and smooth-muscle relaxant properties. Used in peripheral vascular disease. Side effects include flushing, tachycardia, nausea and vomiting.

Tolbutamide
Oral antidiabetic drugs with actions, uses and adverse effects similar to **Chlorpropamide** but excreted more rapidly and thus shorter acting. Recommended when there is greater danger of hypoglycaemia, e.g. in the elderly.

Tolnaftate
Antifungal agent used as cream or powder for skin infections.

Tolu
Balsam obtained from trunk of *Myroxylon balsamum*. Used in cough mixtures for expectorant action and flavour.

Tragacanth
Purgative. Increases faecal bulk by same mechanism as **Methylcellulose**. Occasionally causes allergic rashes or asthma.

Tramazoline
Sympathomimetic agent used topically as nasal decongestant. Actions and adverse effects similar to **Naphazoline**.

Tranexamic acid
Antifibrinolytic agent used to reverse effects of **Streptokinase** or other fibrinolytic activity.

Tranylcypromine
Antidepressant. Inhibits monoamine oxidase. Actions, adverse effects similar to **Phenelzine** but less likely to cause hepatitis. Also has **Amphetamine**-like properties.

Tretinoin
Vitamin A derivative. Used topically in some dermatological treatments, e.g. for acne.

Triamcinolone
Potent synthetic corticosteroid with actions, etc. similar to **Cortisone.** Used mainly for topical treatment of certain skin rashes, e.g. eczema.

Triamterene
Potassium-sparing diuretic similar to **Amiloride.**

Trichlorofluoromethane
Aerosol propellant/refrigerant. Produces intense cold by its rapid evaporation and thus makes tissues insensitive to pain. Used for relief of muscle pain and spasm.

Trichlocarban
Disinfectant used in skin preparations and shampoo for prevention/treatment of certain bacterial and fungal infections. Large doses may cause methaemoglobinaemia.

Triclofos
Hypnotic/sedative. Hydrolysed in stomach to trichloroethanol and absorbed as such. More palatable and causes less gastric irritation than **Chloral hydrate** to which it is otherwise similar.

Triethanolamine
Emulsifying agent used as ear drops to soften wax for removal. May cause localised skin rashes.

Trifluoperazine
Phenothiazine tranquilliser/anti-emetic similar to **Chlorpromazine.**

Trifluperidol
Butyrophenone tranquilliser with actions, uses and adverse effects similar to **Haloperidol.**

Tri-isopropylphenoxy-polyethoxyethanol
Dispersant/emulsifying agent. Used to stabilise oil-in-water mixtures and to disperse and repel spermatozoa thus preventing conception.

Trimeprazine
Phenothiazine similar to **Chlorpromazine.** Used for anti-emetic, sedative and antipruritic effects. Adverse effects, etc. as for chlorpromazine.

Trimetaphan
Antihypertensive with actions and adverse effects similar to **Hexamethonium** but has very brief duration of action. Used for production of controlled hypotension to reduce blood loss during surgery.

Trimethadione (Troxidone)
Anticonvulsant. Suppresses epileptic discharges in the brain. Used in treatment of absence seizures (petit mal) but not for major seizures. May cause sedation, glare phenomenon, photophobia and blood disorders. May precipitate major fits in susceptible patients. Coma with respiratory depression in overdosage. No antidote. Supportive treatment only.

Trimethoprim
Antimicrobial. Inhibits conversion of **Folic acid** to folinic acid. Combined with **Sulphamethoxazole** in **Cotrimoxazole.**

Trimetrazine
Vasodilator used for prevention of angina of effort and intermittent claudication. May cause gastro-intestinal disturbance.

Trimipramine
Antidepressant. Actions and adverse effects similar to **Amitriptyline.**

Tri-potassium di-citrato bismuthate
Antacid with actions of **Bismuth antacids.** Claimed to aid ulcer healing. May cause blackening of tongue and faeces; nausea and vomiting.

Triprolidine
Antihistamine with actions, uses and adverse effects similar to **Promethazine.**

Tropicamide
Parasympatholytic used in the eye as a mydriatic and cycloplegic. Actions, etc. similar to **Atropine** but has rapid onset and short duration of action.

Troxerutin
Vitamin derivative claimed to improve strength and reduce permeability of blood vessels. Used to treat haemorrhoids and venous disorders in the legs.

Troxidone
See **Trimethadone.**

Tryptophan
Amino acid, essential component of diet. Converted in the body to 5-hydroxytryptamine, a neurotransmitter substance that may be depleted in depression. Used in treatment of depression. May cause nausea, drowsiness and may interact with monoamine oxidase inhibitors, e.g. **Phenelzine.**

Tuberculin
Diagnostic agent for tuberculosis. Intradermal injection produces skin reaction in positive cases. May cause skin necrosis in highly sensitive cases.

Tubocurarine
Skeletal muscle relaxant. Blocks passage of impulses at the neuromuscular junction. Used as an adjunct to anaesthesia. May cause fall in blood pressure and paralysis of respiration. **Neostigmine** and **Atropine** plus assisted respiration may be used in treatment of toxic effects.

Turmeric
Yellow colouring agent with mild spicy flavour, a constituent of curry powder. Included in treatment of 'biliary disorders'.

Turpentine oil
Extract of pine used externally as a rubefacient. May cause rashes and vomiting. Rarely used internally but acts as an evacuant if given rectally.

Tyloxapol
Mucolytic. Administered by inhalation from a nebuliser. Liquefies mucus and aids expectoration where viscid mucus is troublesome, e.g. chronic bronchitis. May cause inflammation of eyelids. If left open, the solution is prone to bacterial infections.

Tyrothricin
Antimicrobial. Too toxic for systemic use but used for topical treatment of skin, mouth or ear infections.

U

Undecenoic acid
Antifungal. Applied topically to skin, e.g. in treatment of tinea pedis (athlete's foot).

Uramustine
Cytotoxic drug with actions, uses and adverse effects similar to **Mustine hydrochloride.**

Urea
Osmotic diuretic with actions and uses similar to **Mannitol.** May cause gastric irritation with nausea and vomiting. Intravenous use may cause fall in blood pressure and venous thrombosis at site of injection. Largely superseded by mannitol and other diuretics. Topically in a cream it is used to reduce excess scaling (ichthyosis) and soften the skin.

Urethane
Cytotoxic drug. Used in treatment of certain neoplastic diseases but largely superseded by newer drugs. May cause gastro-intestinal disturbance and bone marrow depression. Has also mild hypnotic properties and is used as an anaesthetic for small anaimals.

Uridine triphosphoric acid
One of the components of ribonucleic acid. Claimed to be of value in treatment of muscular weakness and atrophy.

V

Valerian
C.N.S. depressant occasionally used for sedative effects. No advantage over other drugs that are in more common usage.

Vancomycin
Antibiotic used in infections with penicillin-resistant staphylococci. Must be given intravenously. Adverse effects include ototoxicity.

Vasopressin
Posterior pituitary hormone. Has antidiuretic action on kidney and constricts peripheral blood vessels. Used by injection in diagnosis and treatment of diabetes insipidus. May be used to control bleeding from oesophageal varices. May cause pallor, nausea, eructations, cramps and angina.

Verapamil
Used in prevention of angina of effort and in treatment of cardiac dysrrhythmias. May cause nausea, dizziness and fall in blood pressure. Contraindicated in heart failure.

Veratrum
Natural product. Reduces sympathetic tone. Was used in hypertension, but seldom now because of adverse effects which include nausea, vomiting, sweats, dizziness, respiratory depression and abnormal heart rhythms.

Vidarabine
Antiviral agent with actions, adverse effects similar to **Cytarabine.**

Viloxazine
Antidepressant with some anticonvulsant properties. Unlike the tricyclic antidepressants, e.g. **Imipramine,** it is said to have no anticholinergic or sedative properties. Used in treatment of depression especially when those effects occur from other drugs. May cause nausea and vomiting. If given with **Phenytoin** may induce toxicity due to that drug. No antidote. Overdosage treated symptomatically.

Vinblastine
Cytotoxic alkaloid from West Indian periwinkle, used in neoplastic disease. Adverse effects include neuropathy and bone marrow depression.

Vincristine
Cytotoxic with actions and adverse effects as for **Vinblastine.**

Viomycin
Antibiotic with actions and adverse effects similar to **Streptomycin.**

Viprynium
Used in treatment of threadworms. Adverse effects include red stools, vomiting, diarrhoea.

Vitamin A
Fat-soluble vitamin present in liver, dairy products and some vegetables, essential for normal visual function and for maintenance of epithelial surfaces. Overdosage produces mental changes, hyperkeratosis and hypoprothrombinaemia.

Vitamin B_1
See **Aneurine.**

Vitamin B_2
See **Riboflavine.**

Vitamin B_6
See **Pyridoxine.**

Vitamin B_7
See **Nicotinic acid.**

Vitamin B_{12}
See **Hydroxocobalamin.**

Vitamin C (Ascorbic acid)
Vitamin found in fruit and vegetables, necessary for normal collagen formation. Deficiency causes scurvy with mucosal bleeding and anaemia. High-dose administration in prophylaxis against common cold is controversial.

Vitamin D (Calciferol)
Group of fat-soluble vitamins found in dairy products and formed in skin exposed to sunlight. Promotes gut absorption of calcium and its mobilisation from bone. Deficiency produces rickets and bone softening. Excess produces hypercalcaemia, ectopic calcification and renal failure.

Vitamin E (Tocopheryl)
Vitamin with no clearly defined requirements or deficiency disease in man. Has been suggested as treatment for habitual abortion, cardiovascular disease and other conditions.

Vitamin K (Menaphthone, Menadiol, Phytomenadione, Acetomenaphthone)
Fat-soluble vitamin responsible for formation of prothrombin and other clotting factors. Used to reverse oral anticoagulants and in bleeding diseases. Excessive dosing may produce haemolysis.

W X Y

Warfarin
Coumarin anticoagulant which interferes with synthesis of clotting factors by the liver. May produce allergic reactions. Overdosage produces haemorrhage controlled by **Vitamin K.** Potentiated by drugs such as **Acetysalicylic acid** and **Phenylbutazone** which displace from protein binding, and reduced by hepatic enzyme inducers such as barbiturates. Caution if used in liver disease.

Xanthinol nicotinate
Peripheral vasodilator with actions similar to **Nicotinic acid.** Claimed to be of value in treating reduced peripheral and cerebral circulation. May cause flushing, hypotension and abdominal pain.

Xylometazoline
Sympathomimetic used topically as nasal decongestant. Actions and adverse effects similar to **Naphazoline.**

Yohimbine
Plant extract with alpha adrenoceptor-blocking actions. Said to have aphrodisiac properties but not proven.

Z

Zinc chloride
Astringent and deodorant. Used in mouth wash and for application to wounds. Caustic in higher concentrations.

Zinc Ichthammol
Mixture of **Zinc oxide** and **Ichthammol** used for treatment of eczema.

Zinc oleate
Topical treatment for eczema. Actions similar to **Zinc oxide.**

Zinc oxide
Dermatological treatment. Has mild astringent, soothing and protective effects. Used topically in treatment of eczema and excoriated skin.

Zinc phenolsulphonate
Antiseptic dusting powder similar to **Zinc oxide.**

Zinc powder
See **Zinc oxide.**

Zinc salicylate
Dermatological treatment with actions and uses similar to **Zinc oxide.**

Zinc sulphate
Astringent used topically for skin wounds/ulcers to assist healing. Also included in some eye drops for minor allergic conjunctivitis. Orally it is an emetic but is not used for this purpose due to toxic effects.

Zinc undecenoate
Antifungal used topically for fungal infections of the skin.

Warwickshire School of Nursing
Nurse Education Centre
Central Hospital
Near Warwick, Warwickshire CV35 7EB

Warwickshire School of Nursing
Nurse Education Centre
Central Hospital
Near Warwick, Warwickshire CV35 7EB

Part II

Trade Names Index

A

AAA Mouth and Throat Spray.   Local anaesthetic/antiseptic for sore throat:
see **Benzocaine, Cetalkonium.**
Abacid (d).   Antacid: see **Magnesium trisilicate, Aluminium hydroxide.**
Abacid plus (d).   Antacid/anticholinergic: as Abacid plus **Dicyclomine,
Dimethicone.**
Abecedin (d).   Compound vitamins: see **Aneurine, Riboflavin, Vitamin C,
Vitamin D.**
Abicol.   Antihypertensive: see **Reserpine, Bendrofluazide.**
Abidec.   Vitamin mixture: see **Vitamin A, Vitamin D, Aneurine, Riboflavine.**
Abstem.   Treatment of alcoholism: see **Calcium carbimide.**
Achromycin.   Anti-infective.   See **Tetracycline.**
Acidol-Pepsin.   Acid/Pepsin mixture for use in achlorhydria.
Aci-Jel.   Jelly for topical treatment of vaginal infection.
Acnil.   Topical treatment for acne: see **Resorcinol, Cetrimide.**
Actal.   Antacid: see **Aluminium carbonate.**
Acthar gel.   Corticotrophic hormone injection: see **Corticotrophin.**
Actidil.   Antihistamine: see **Triprolidine.**
Actifed.   Decongestant: see **Triprolidine, Pseudoephedrine.**
Actifed Compound Linctus.   As Actifed plus **Codeine** for cough suppression.
Actinac.   Topical treatment for acne: see **Chloramphenicol, Hydrocortisone.**
Actonorm.   Antacid: see **Aluminium hydroxide, Magnesium hydroxide,
Simethicone.**
Actonorm-Sed.   As Actonorm plus **Atropine sulphate** and **Phenobarbitone.**
Adabee (d).   Vitamin supplements: see **Aneurine, Riboflavin, Vitamin C,
Pyridoxine, Nicotinic acid.**
Adcortyl.   Corticosteroid for topical or systemic use: see **Triamcinalone.**
Adelphane (d).   Antihypertensive: see **Reserpine, Hydrallazine.**
Adenotriphos.   For cardiovascular disease: see **Adenosine triphosphoric
acid (ATP).**
Adenyl.   Non-steroid anti-inflammatory: see **Adenosine monophosphoric
acid (AMP).**
Adexolin (d).   Vitamin supplement: see **Vitamin A, Vitamin D.**
Admune.   Inactivated influenza virus vaccine.
Adriamycin.   Cytotoxic antibiotic: see **Doxorubicin.**
Aerosporin.   Antibiotic: see **Polymyxin B.**
Agarol.   Laxative: see **Liquid paraffin, Phenolphthalein.**
Aglutella Gentili (b).   **Gluten**-free pasta, low in protein, sodium and
potassium.   Dietary substitute used in chronic renal failure and phenyl-
ketonuria.
Agocholine (d).   Laxative: see **Magnesium sulphate.**
Airbron.   Mucolytic: see **Acetylcysteine.**
Akineton.   Anticholinergic, anti-Parkinsonian: see **Biperiden.**

Akrotherm.   Topical treatment for chilblains: see **Histamine, Acetylcholine.**
Alavac.   Vaccines for desensitisation to pollens and other allergens in asthma
and hay fever.
Albamycin.   Antibiotic: see **Novobiocin.**
Albamycin GU.   Urinary anti-infective: see **Novobiocin, Sulphamethizole.**
Albamycin-T.   Antibiotic: see **Novobiocin, Tetracycline.**
Albucid.   Anti-infective eye drops: see **Sulphacetamide.**
Albumaid Preps (b).   Range of dietary substitutes for use in malabsorption
and inherited metabolic disorders, e.g. phenylketonuria.
Alcin.   Antacid: see **Magnesium** and **Aluminium salts.**
Alcopar.   Anthelmintic: see **Bephenium.**
Alcos-Anal.   Local anaesthetic for haemorrhoids: see **Benzocaine, Sodium
morrhuate, Chlorothymol.**
Aldactide.   Diuretic/antihypertensive: see **Spironolactone, Hydroflumethia-
zide.**
Aldactone-A.   Diuretic: see **Spironolactone.**
Aldocorten.   Sodium-retaining adrenocorticosteroid: see **Aldosterone.**
Aldomet.   Antihypertensive: see **Methyldopa.**
Alembicol D (b).   Lipid extract of coconut oil; substitute for long-chain
fats in fat malabsorption.
Aleudrin.   Sympathomimetic bronchodilator: see **Isoprenaline.**
Alevaire.   Mucolytic: see **Tyloxapol.**
Algesal.   Rubefacient: see **Diethylamine salicylate.**
Algipan.   Rubefacient: see **Salicylic acid, Nicotinic acid, Histamine, Capsicum.**
Alka-Donna (d).   Antacid: see **Magnesium trisilicate, Aluminium hydroxide,
Belladonna extract.**
Alka-Donna P (d).   Antacid/sedative as Alka-Donna plus **Phenobarbitone.**
Alkeran. Cytotoxic: see **Melphalan.**
Allbee with C.   Vitamin supplement: see **Aneurine, Riboflavin, Pyridoxine,
Nicotinic acid, Vitamin C, Pantothenic acid.**
Allegron.   Antidepressant: see **Nortriptyline.**
Alloferin.   Muscle relaxant: see **Alcuronium.**
Allpyral.   Desensitising vaccines similar to Alavac.
Almacarb (d).   Antacid: see **Aluminium hydroxide, Magnesium carbonate.**
Almevax.   Live attenuated rubella (German measles) virus vaccine.
Alophen.   Purgative: see **Aloin, Phenolpthalein, Ipecacuanha, Belladonna
extract.**
Alphaderm.   Topical corticosteroid cream: see **Hydrocortisone.**
Alphosyl.   Topical treatments for psoriasis and other scaly disorders: see
Allantoin, Coal tar.
Alphosyl HC.   Topical steroid treatment for psoriasis: see **Allantoin, Coal
tar, Hydrocortisone.**
Alrheumat.   Non-steroid anti-inflammatory: see **Ketoprofen.**
Altacite.   Antacid: see **Hydrotalcite.**
Althesin.   Short-acting anaesthetic used for induction of anaesthesia or alone
for short operations: see **Alphaxalone, Alphadolone.**
Alubarb.   Antacid/sedative: see **Aluminium hydroxide, Phenobarbitone,
Belladonna extract.**

Alu-Cap. Antacid: see **Aluminium hydroxide.**

Aludrox. Antacid: see **Aluminium hydroxide.**

Aludrox SA. Antacid/sedative: see **Aluminium hydroxide, Secbutobarbitone, Ambutonium.**

Aluhyde (d). Antacid/sedative: see **Aluminium hydroxide, Magnesium trisilicate, Quinalbarbitone, Belladonna extract.**

Alupent. Sympathomimetic bronchodilator: see **Orciprenaline.**

Alupent expectorant. Bronchodilator/mucolytic: see **Orciprenaline, Bromhexine.**

Alupent-Sed (d). Sympathomimetic bronchodilator/sedative: see **Orciprenaline, Amylobarbitone.**

Aluphos. Antacid: see **Aluminium antacids.**

Aluzyme. Vitamin mixture: see **Aneurine, Riboflavin, Nicotinic acid, Folic acid.**

Amargyl (d). Sedative/tranquilliser: see **Chlorpromazine, Amylobarbitone.**

Ambilhar (d). Anthelmintic: see **Niridazole.**

Amenorone (d). Synthetic gonadal hormones for symptomatic treatment of amenorrhoea: see **Ethinyloestradiol, Ethisterone.**

Amesac. Bronchodilator/sedative: see **Aminophylline, Ephedrine, Amylobarbitone.**

Amfac (d). Fraction of liver extract used to stop functional uterine haemorrhage.

Amfipen. Antibiotic: see **Ampicillin.**

Amikin. Antibiotic: see **Amikacin.**

Amin-ex biscuits (b). Low amino acids/protein dietary substitute for use in phenylketonuria and renal failure.

Aminofusin L600 and L1000. Amino acid solutions for intravenous feeding.

Aminogran (b). Low amino acid food substitute for phenylketonuria.

Aminomed. Bronchodilator: see **Theophylline.**

Aminomed Compound. Bronchodilator: see **Theophylline, Ephedrine.**

Aminoplex 5 and 14. Intravenous amino acids, **Sorbitol,** ethanol, vitamins and electrolytes.

Aminosol. Intravenous amino acids, peptides and electrolytes.

Aminutrin (b). Protein source for tube feeding.

Amisyn. Peripheral vasodilator: see **Acetomenaphthone, Nicotinamide.**

Amoxil. Antibiotic: see **Amoxycillin.**

Ampiclox. Antibiotic: see **Ampicillin, Cloxacillin.**

Amplex (b). Oral deodorant.

Amplexol (b). Topical deodorant.

Amylomet. Hypnotic/sedative plus emetic to prevent absorption in overdosage: see **Amylobarbitone, Emetine.**

Amylozine spansule. Slow-release hypnotic, minor tranquilliser: see **Amylobarbitone.**

Amytal. Hypnotic, minor tranquilliser: see **Amylobarbitone.**

Anabolex. Anabolic steroid: see **Stanolone.**

Anacal. Topical treatment for haemorrhoids: see **Prednisolone, Hexachlorophane.**

Anaflex lozenges. Antiseptic for mouth and throat: see **Polynoxylin.**

Anaflex topical preps.   Skin antiseptic: see **Polynoxylin.**
Anafranil.   Antidepressant: see **Clomipramine.**
Anahaemin (d).   Liver fraction for pernicious anaemia: see **Cyanocobalamin.**
Analgin (d).   Analgesic: see **Acetylsalicylic acid, Phenacetin, Codeine, Caffeine.**
Ananase forte.   Proteolytic enzymes for post-traumatic tissue reactions.
Anapolon.   Anabolic steroid: see **Oxymetholone.**
Ancoloxin.   Anti-emetic: see **Meclozine, Pyridoxine.**
Androcur.   For male sexual disorders: see **Cyproterone.**
Andursil.   Antacid: see **Aluminium hydroxide, Magnesium hydroxide, Magnesium carbonate.**
Anectine.   Muscle relaxant during anaesthesia: see **Suxamethonium.**
Anethaine.   Topical anaesthetic for haemorrhoids: see **Amethocaine.**
Anodesyn.   Local treatment for haemorrhoids: see **Ephedrine, Lignocaine, Allantoin, Bronopol.**
Anorvit.   For anaemia: see **Ferrous sulphate, Vitamin C, Acetomenaphthone.**
Anovlar 21.   Oral contraceptive: see **Ethinyloestradiol, Norethisterone.**
Anquil.   Tranquilliser: see **Benperidol.**
Ansolysen (d).   Antihypertensive: see **Pentolinium.**
Antabuse 200.   For treatment of alcoholism: see **Disulfiram.**
Antasil.   Antacid: see **Aluminium hydroxide, Magnesium hydroxide, Dimethicone.**
Antemin (d).   Spermicidal cream: see **Spermicides.**
Antepar.   For threadworms and roundworms: see **Piperazine.**
Anthical.   Antihistamine cream for skin inflammation: see **Mepyramine, Zinc oxide.** May cause skin rash.
Anthiphen.   For tapeworms: see **Dichlorophen.**
Anthisan cream.   Antihistamine for insect bites: see **Mepyramine.**
Anthisan elixir.   Antihistamine for allergy: see **Mepyramine.**
Antidiar.   Antidiarrhoeal: see **Aluminium hydroxide.**
Antidiar with Neomycin.   Antidiarrhoeal/antibacterial: see   **Aluminium hydroxide, Neomycin.**
Antidol.   Analgesic: see **Ethosalmide, Paracetamol, Caffeine.**
Antisin-Privine.   Anti-allergic nasal decongestant: see **Antazoline, Naphazoline.**
Antoin.   Analgesic: see **Acetylsalicylic acid, Codeine, Caffeine, Calcium carbonate.**
Antrenyl. (d)   Anticholinergic for peptic ulcers: see **Oxyphenonium.**
Anturan.   Increases urate excretion in gout: see **Sulphinpyrazone.**
Anugesic-HC.   Topical treatment for haemorrhoids: see **Pramoxine, Hydrocortisone, Zinc oxide, Benzyl benzoate.**
Anusol.   Local treatment for haemorrhoids: see **Zinc oxide, Benzyl benzoate, Bismuth subgallate, Bismuth oxide.**
Anusol HC.   Local treatment for haemorrhoids: see **Anusol, Hydrocortisone.**
Apisate.   Anti-obesity: see **Diethylpropion, Thiamine.**
APP.   Antacid/anticholinergic for gastro-intestinal disorders: see **Homatropine, Papaverine, Phenobarbitone, Calcium carbonate, Magnesium carbonate, Aluminium hydroxide.**

Apresoline.   Antihypertensive: see **Hydrallazine.**
Aprinox.   Diuretic: see **Bendrofluazine.**
Aproten (b).   **Gluten**-free products for coeliac disease.
Apsin V.K.   Antibiotic: see **Phenoxymethyl penicillin.**
Aquadrate.   Keratolytic for thickened, dry skin: see **Urea.**
Aquamephyton.   For prothrombin deficiency: see **Phytomenadione.**
Aquamox.   Diuretic: see **Quinethazone.**
Aradolene.   Rubefacient: see **Salicylate.**
Aramine.   Vasoconstrictor for shock: see **Metaraminol.**
Ardena (b).   Concealing cream: see **Zinc oxide, Titanium dioxide.**
Arfonad.   Hypotensive: see **Trimetaphan.**
Arginine-Sorbitol EGIC.   Intravenous infusion for liver coma: see **Arginine, Sorbitol.**
Argotone.   Nasal decongestant: see **Ephedrine, Silver Protein.**
Arlef.   Non-steroid anti-inflammatory/analgesic: see **Flufenamic acid.**
Artane.   Antiparkinsonian: see **Benzhexol.**
Arvin.   Anticoagulant: see **Ancrod.**
Ascabiol.   Topical treatment for scabies and lice: see **Benzyl benzoate.**
Ascon.   Antacid/antispasmodic for peptic ulcers: see **Aluminium hydroxide, Magnesium trisilicate, Hyoscyamine.**
Aserbine.   Topical treatment for skin ulcers and wounds: see **Malic acid, Benzoic acid, Salicylic acid.**
Asilone.   Antacid: see **Dimethicone, Aluminium hydroxide.**
Asmacort (d).   For asthma: see **Prednisone, Ephedrine, Theophylline, Phenobarbitone.**
Asmal.   For asthma: see **Ephedrine, Theophylline, Phenobarbitone.**
Asmapax.   For asthma: see **Ephedrine, Theophylline, Bromvaletone.**
Asma-Vydrin.   Inhalation for asthma: see **Adrenaline, Atropine methonitrate, Papaverine.**
Aspellin.   Rubefacient: see **Acetylsalicylic acid, Methyl salicylate.**
Aspergum.   Analgesic/antipyretic chewing gum: see **Acetylsalicylic acid.**
Asterol (d).   Topical treatment for fungal skin conditions: see **Diamthazole.**
Astiban.   For treatment of schistosomiasis: see **Stibocaptate.**
A.T. 10.   For Vitamin D deficiency or resistance: see **Dihydrotachysterol.**
Atarax.   Sedative/tranquilliser: see **Hydroxyzine.**
Atasorb.   Antidiarrhoeal: see **Attapulgite.**
Atasorb-N.   Antidiarrhoeal/antibacterial: see **Attapulgite, Neomycin.**
Atensine.   Sedative/tranquilliser: see **Diazepam.**
Ativan.   Sedative/tranquilliser: see **Lorazepam.**
Atromid-S 500.   Anti-anginal/lipid-lowering agent: see **Clofibrate.**
Audax.   Drops for middle and outer ear infection: see **Choline salicylate.**
Audicort.   Drops for outer ear infection: see **Neomycin, Triamcinolone, Benzocaine.**
Auralgicin.   Drops for middle ear pain: see **Ephedrine, Benzocaine, Phenazone.**
Auraltone.   Drops for middle ear pain: see **Benzocaine, Phenazone.**
Aureocort.   Topical treatment for allergic/infective skin conditions: see **Chlortetracycline, Triamcinolone.**

Aureomycin.   Antibiotic: see **Chlortetracycline.**
Aveeno.   For addition to bath for skin allergy.
Aventyl.   Antidepressant: see **Nortriptyline.**
Avloclor.   Antimalarial: see **Chloroquine.**
Avomine.   Antihistamine/anti-emetic: see **Promethazine.**
Azeta (b).   Low-protein biscuits for renal failure, phenylketonuria.

B

Bactrim.   Antibacterial: see **Cotrimoxazole.**
Balmosa.   Rubefacient: see **Methyl salicylate, Benzyl nicotinate.**
Banistyl.   Antihistamine: see **Dimethothiazine.**
Banocide.   For filariasis: see **Diethylcarbamazine.**
Barquinol HC.   Topical corticosteroid cream for allergic/infectious skin disorders: see **Hydrocortisone, Clioquinol.**
Baycaron.   Diuretic: see **Mefruside.**
Bayolin.   Rubefacient: see **Glycol salicylate.**
B.C. 500.   Vitamin mixture: see **Vitamin C, Thiamine, Riboflavine, Nicotinamide, Pyridoxine, Pantothenic acid, Cyanocobalamin.**
Beconase.   Nasal aerosol for allergic rhinitis: see **Beclomethasone.**
Becosed.   Sedative/tranquilliser: see **Phenobarbitone, Thiamine, Riboflavine, Nicotinamide.**
Becosym.   Vitamin B complex: see **Thiamine, Riboflavine, Nicotinamide, Pyridoxine.**
Becotide.   Steroid aerosol for asthma: see **Beclomethasone.**
Beflavit (d).   Vitamin: see **Riboflavine.**
Belladenal.   For reducing gastro-intestinal motility and secretion: see **Belladonna, Phenobarbitone.**
Bellergal.   Sedative/tranquilliser: see **Belladonna, Ergotamine, Phenobarbitone.**
Bellobarb.   For reducing gastro-intestinal motility and secretion: see **Belladonna, Phenobarbitone, Magnesium trisilicate.**
Benadon.   Anti-emetic: see **Pyridoxine.**
Benadryl.   Antihistamine for allergy: see **Diphenhydramine.**
Benafed.   Cough suppressant/decongestant: see **Diphenhydramine, Dextromorphan, Pseudoephedrine.**
Benemid.   Uricosuric for gout: see **Probenecid.**
Benerva.   Vitamin for beri-beri and neuritis: see **Thiamine.**
Benerva compound.   Vitamin B mixture for deficiency: see **Thiamine, Riboflavine, Nicotinamide.**
Bengue's balsam.   Rubefacient: see **Methyl salicylate.**
Benoral. Non-steroid anti-inflammatory/analgesic: see **Benorylate.**
Benoxyl.   Topical treatment for acne: see **Benzoyl peroxide.**
Benuride.   Anticonvulsant: see **Pheneturide.**
Benylin.   Cough suppressant: see **Diphenhydramine.**

Benylin with codeine. Cough suppressant: see **Diphenhydramine, Codeine.**
Benzedrex inhaler. Nasal decongestant: see **Propylhexedrine.**
Benzets. Lozenges for mouth and throat infections: see **Benzalkonium, Benzocaine.**
Benogex. Purgative: see **Sodium phosphate.**
Beplete. Sedative/tranquilliser with vitamins: see **Phenobarbitone, Thiamine, Riboflavine, Pyridoxine, Nicotinamide.**
Berkdopa. Antiparkinsonian: see **Laevodopa.**
Berkfurin. Urinary anti-infective: see **Nitrofurantoin.**
Berkmycen. Antibiotic: see **Oxytetracycline.**
Berkomine. Antidepressant: see **Imipramine.**
Berkozide. Diuretic: see **Bendrofluazide.**
Berotec. Bronchodilator for asthma: see **Fenoterol.**
Beta-Cardone. Beta adrenoceptor blocker: see **Sotalol.**
Betadine. Antiseptic: see **Povidone-iodine.**
Betaloc. Beta adrenoceptor blocker: see **Metroprolol.**
Betnelan. Corticosteroid: see **Betamethasone.**
Betnesol. Soluble corticosteroid tablets and injection: see **Betamethasone.**
Betnesol-N. Topical corticosteroid/antibiotic: see **Betamethasone, Neomycin.**
Betnovate. Topical corticosteroid: see **Betamethasone.**
Betnovate A. Topical corticosteroid/antibiotic: see **Betamethasone, Chlortetracycline.**
Betnovate C. Topical corticosteroid/anti-infective: see **Betamethasone, Clioquinol.**
Betnovate N. Topical corticosteroid/antibiotic: see **Betamethasone, Neomycin.**
Bextasol inhaler. Steroid aerosol for asthma: see **Betamethasone.**
B.F.I. Topical anti-infective powder: see **Bismuth-formic-iodide, Zinc phenolsulphonate, Bismuth subgallate, Boric acid.**
Bi-Aglut (b). **Gluten**-free biscuits for gluten-sensitive bowel disorders.
Bicillin. Antibiotic: see **Procaine penicillin, Benzylpenicillin.**
Biogastrone. For gastric ulcers: see **Carbenoxolone.**
Biomydrin. Nasal decongestant/anti-inflammatory: see **Neomycin, Gramicidin, Thonzylamine, Phenylephrine.**
Bioral. Topical therapy for mouth ulcers: see **Carbenoxolone.**
Bislumina suspension. Antacid: see **Bismuth aluminate.**
Bisolvomycin. Mucolytic/antibiotic: see **Bromhexine, Oxytetracycline.**
Bisolvon. Mucolytic: see **Bromhexine.**
Blocadren. Beta adrenoceptor blocker: see **Timolol.**
Bocasan. Antiseptic mouthwash for oral infections: see **Sodium perborate.**
Bolvidon. Antidepressant: see **Mianserin.**
Bonjela. Topical therapy for mouth ulcers: see **Choline salicylate, Cetalkonium.**
Bradilan. Peripheral vasodilator/lipid-lowering agent: see **Tetranicontinoylfructose.**
Bradosol. Antiseptic for mouth and throat infections: see **Domiphen.**
Brasivol. Topical abrasive/cleansing paste for acne: see **Aluminium oxide.**

Bravit.　Vitamin mixture: see **Aneurine, Riboflavine, Pyridoxine, Nicotinic acid, Vitamin C.**
Breoprin.　Sustained-release analgesic: see **Acetylsalicylic acid.**
Bretylate.　Antidysrhythmic: see **Bretylium.**
Brevidil.　Muscle relaxant used during anaesthesia: see **Suxamethonium.**
Brevinor.　Oral contraceptive: see **Ethinyloestradiol, Norethisterone.**
Bricanyl.　Sympathomimetic bronchodilator: see **Terbutaline.**
Bricanyl expectorant.　Bronchodilator/expectorant.　As Bricanyl plus **Guaiphenesin.**
Brietal Sodium.　Short-acting barbiturate for short-duration anaesthesia: see **Methohexitone.**
Brinaldix.　Diuretic: see **Clopamide.**
Brinaldix K.　As Brinaldix plus **Potassium chloride** supplement.
Brizin.　Anticholinergic/antiparkinsonian: see **Benapryzine.**
Brocadopa.　Antiparkinsonian: see **Laevodopa.**
Brocadopa Temtabs.　Sustained-release formulation of Brocadopa.
Brolene ophthalmic preps.　Anti-infective drops/ointment for use in the eye: see **Propamidine.**
Bronchilator.　Aerosol bronchodilator: see **Isoetharine, Phenylephrine, Thenyldiamine.**
Bronchotone.　Bronchodilator syrup: see **Ephedrine, Sodium salicylate, Belladonna extract, Sodium iodide.**
Brontina.　Bronchodilator: see **Deptropine.**
Brontisol.　Aerosol bronchodilator: see **Deptropine, Isoprenaline.**
Brontyl 300.　Bronchodilator: see **Proxyphylline.**
Brovon inhalant.　Aerosol bronchodilator: see **Adrenaline, Atropine methonitrate, Papaverine.**
Broxil.　Antibiotic: see **Phenethicillin.**
Brufen.　Non-steroid anti-inflammatory/analgesic: see **Ibuprofen.**
Brulidine.　Topical anti-infective for burns, wounds: see **Dibromopropamidine.**
Budale.　Analgesic/sedative: see **Paracetamol, Codeine, Butobarbitone.**
Bufferin (d).　Analgesic: see **Acetylsalicylic acid** plus **Aluminium glycinate** to reduce gastric irritation.
Burinex.　Diuretic: see **Bumetanide.**
Burinex K.　As Burinex plus **Potassium chloride** supplement.
Buscopan.　Anticholinergic/antispasmodic for gastro-intestinal or uterine spasm: see **Hyoscine butylbromide.**
Butacote.　Non-steroid anti-inflammatory/analgesic: see **Phenylbutazone.** Enteric-coated to reduce gastric irritation.
Butazolidin.　Non-steroid anti-inflammatory/analgesic: see **Phenylbutazone.**
Butazolidin Alka.　As Butazolidin plus antacids, **Aluminium hydroxide, Magnesium trisilicate,** to reduce gastric irritation.
Butazone.　Non-steroid anti-inflammatory/analgesic: see **Phenylbutazone.**
Butomet.　Hypnotic/sedative plus emetic to reduce absorption: see **Butobarbitone, Emetine.**

C

Cafadol.   Analgesic/antipyretic: see **Paracetamol, Caffeine.**

Cafergot.   Vasoconstrictor for migraine: see **Ergotamine, Caffeine, Belladonna alkaloids, Allylbarbituric acid.**

Caladryl.   Cream or lotion for skin irritation, e.g. sunburn, insect bites: see **Calamine, Diphenhydramine.**

Calcidrine (d).   Cough linctus: see **Calcium iodide, Ephedrine, Pentobarbitone, Codeine.**

Calcimax.   Calcium/vitamin supplements for calcium deficiencies: see **Calcium, Vitamin D, Aneurine, Riboflavine, Pyridoxine, Hydroxocobalamin, Vitamin C, Nicotinic acid, Pantothenic acid.**

Calcitare.   Hormone: see **Calcitonin.**

Calcium heparin.   Anticoagulant: see **Heparin.**

Calcium Resonium.   Ion-exchange resin: see **Calcium polystyrene sulphonate.**

Calcium-Sandoz.   Calcium supplement for deficiency, e.g. tetany.

Callusolve.   Topical treatment for warts: see **Benzalkonium.**

Calmoden.   Anxiolytic: see **Chlordiazepoxide.**

Calmurid.   Keratolytic cream for removal of dry, scaly skin: see **Urea.**

Calmurid HC.   Corticosteroid cream for eczema: see **Urea, Hydrocortisone.**

Calonutrin (b).   Mono-, di- and poly-saccharide mixture free from lactose and sucrose.

Caloreen (b).   Protein-free, high-calorie powder for use when low-protein diet is needed, e.g. kidney failure.

Calpol.   Analgesic elixir: see **Paracetamol.**

Calsept (d).   Soothing lotion for sunburn or nappy rash: see **Calamine, Cetrimide, Zinc oxide.**

Calsynar.   Hormone: see **Calcitonin.**

C.A.M.   Bronchodilator elixir for children: see **Ephedrine, Butethamate.**

Camcolit.   Antidepressant for manic-depressive psychosis: see **Lithium salts.**

Candeptin.   Antifungal: see **Candicidin.**

Canesten.   Antifungal: see **Clotrimazole.**

Cantil.   Anticholinergic for abdominal colic and diarrhoea: see **Mepenzolate.**

Cantil with phenobarbitone.   As Cantil plus **Phenobarbitone** as sedative.

Capastat.   Antituberculous antibiotic: see **Capreomycin.**

Capitol (b).   Shampoo for scaly scalp conditions: see **Benzalkonium.**

Caprin.   Slow-release analgesic: see **Acetylsalicylic acid.**

Capsolin.   Rubefacient: see **Capsicum, Camphor, Turpentine oil, Eucalyptus.**

Carbellon.   Antacid/sedative: see **Magnesium hydroxide, Belladonna extract, Charcoal.**

Carbocaine.   Local anaesthetic: see **Mepivacaine.**

Carbo-Cort.   Corticosteroid cream for eczema and other skin rashes: see **Hydrocortisone, Coal tar.**

Carbo-Dome.   Topical treatment for psoriasis: see **Coal tar.**

Carbomucil.   Antidiarrhoeal: see **Charcoal, Sterculia, Magnesium carbonate.**

5*

Carbrital. Hypnotic: see **Pentobarbitone, Carbromal.**

Cardiacap. Antianginal: see **Pentaerythritol tetranitrate.**

Cardiacap-A. As Cardiacap plus **Amylobarbitone** as sedative.

Cardophylin. Bronchodilator/anti-anginal: see **Aminophylline.**

Carisoma. Muscle relaxant: see **Carisprodol.**

Carisoma compound. As Carisoma plus analgesic: see **Carisprodol, Paracetamol, Caffeine.**

Carylderm. Insecticide lotion shampoo for treatment of lice: see **Carbanyl.**

Cascara evacuant. Purgative: see **Cascara.**

Casilan (b). High-protein, low-salt food for hypoproteinaemia.

Catapres. Antihypertensive: see **Clonidine.**

Caved-S. For peptic ulcers: see **Deglycyrrhizinised liquorice, Aluminium hydroxide, Bismuth salts, Sodium bicarbonate, Magnesium carbonate, Frangula.**

Ceanel (b). Shampoo for psoriasis: see **Cetrimide.**

Ce-Cobalin. Vitamins: see **Ascorbic acid, Cyanocobalamin.**

Cedilanid. For heart failure: see **Lanatoside C.**

Cedocard. Anti-anginal: see **Sorbide nitrate.**

Ceduran. Urinary anti-infective: see **Nitrofurantoin, Deglycyrrhizinised liquorice.**

Celbenin. Antibiotic: see **Methicillin.**

Celevac. Purgative: see **Methylcellulose.**

Cellucon. Purgative: see **Methylcellulose.**

Celontin. Anticonvulsant: see **Methsuximide.**

Cendevax. Rubella (German measles) virus vaccine.

Centyl-K. Diuretic: see **Bendrofluazide, Potassium chloride.**

Ceporex. Antibiotic: see **Cephalexin.**

Ceporin. Antibiotic: see **Cephaloridine.**

Cerubidin. Cytotoxic: see **Daunorubicin.**

Cerumol. Drops for removal of ear wax: see **Paradichlorobenzene, Chlorbutol, Turpentine oil.**

Cetavlex. Topical anti-infective for minor abrasions: see **Cetrimide.**

Cetavlon PC (b). Shampoo for seborrhoea: see **Cetrimide.**

Cetavlon powder. Anti-infective powder for skin: see **Cetrimide.**

Cetiprin. Anticholinergic used for relief of post-operative bladder pain and urinary frequency: see **Emepronium.**

Chloractil. Major tranquilliser: see **Chlorpromazine.**

Chloromycetin. Antibiotic for topical or systemic therapy: see **Chloramphenicol.**

Chloromycetin hydrocortisone. Anti-infective/and corticosteroid drops for infection/inflammation of eyes: see **Chloramphenicol/Hydrocortisone.**

Chocovite. For calcium deficiency: see **Calcium gluconate, Vitamin D.**

Choledyl. Bronchodilator: see **Choline theophyllinate.**

Choloxin. Blood cholesterol-lowering agent: see **D-Thyroxine.**

Chu-Pax (d). Chewable analgesic tablets: see **Acetylsalicylic acid.**

Chymacort. Topical anti-infective corticosteroid and proteolytic enzymes for certain skin conditions, e.g. varicose ulcers: see **Neomycin, Hydrocortisone, Pancreatic enzymes.**

Chymar. Enzyme injection for post-traumatic tissue reactions: see **Chymotrypsin.**

Chymar-Zon. Enzyme extract used to aid extraction of the lens from the eye in cataract surgery: see **Chymotrypsin.**

Chymocyclar. Antibiotic plus enzymes: see **Tetracycline, Pancreatic enzymes.**

Chymoral. Oral enzyme preparation for post-traumatic tissue reactions: see **Pancreatic enzymes.**

CIBA-1906. Anti-leprosy: see **Thiambutosine.**

Cicatrin. Topical anti-infective: see **Neomycin, Bacitracin.**

Cidomycin. Antibiotic for injection or topical use: see **Gentamicin.**

Citanest. Local anaesthetic for minor surgery: see **Prilocaine.**

Clairvan. Respiratory stimulant: see **Ethamivan.**

Claradin. Effervescent analgesic: see **Acetylsalicylic acid.**

Climatone. Sex hormones for use in menopause, dysmenorrhoea, premenstrual syndrome, osteoporosis: see **Ethinyloestradiol, Methyltestosterone.**

Clinimycin. Antibiotic: see **Oxytetracycline.**

Clorevan. Antihistamine used in Parkinsonism: see **Chlorphenoxamine.**

Cobadex. Corticosteroid ointment for dermatitis: see **Hydrocortisone.**

Cobalin-H. Vitamin B_{12} injection for treatment of B_{12}-deficient anaemia: see **Hydroxocobalamin.**

Codelcortone. Corticosteroid: see **Prednisolone.**

Codelcortone-TBA. Long-acting corticosteroid injection for intra-articular or soft-tissue injection: see **Prednisolone.**

Codelsol. Corticosteroid injection: see **Prednisolone.**

Codis. Soluble analgesic: see **Acetylsalicylic acid, Codeine.**

Co-Ferol. Prophylactic treatment for anaemia of pregnancy: see **Ferrous fumarate, Folic acid.**

Cogentin. Anticholinergic/antiparkinsonian: see **Benztropine.**

Colbenemid. For gout: see **Probenecid, Colchicine.**

Coliacron. Enzyme preparation (glutamine synthetase and acetyl CoA-kinase) for certain recommended psychosomatic disorders.

Colifoam. Corticosteroid in aerosol foam for topical treatment of inflammation of large bowel: see **Hydrocortisone.**

Colliron (d). Iron-containing syrup for treatment of iron-deficiency anaemia: see **Ferric hydroxide.**

Colofac. Antispasmodic for abdominal colic: see **Mebeverine.**

Cologel. Purgative: see **Methylcellulose.**

Colomycin. Anti-infective: see **Colistin.**

Combizym. For use in pancreatic deficiency: see **Pancreatic enzymes.**

Combizym Composition. For use in pancreatic deficiency: see **Pancreatic enzymes, Bile salts.**

Comminuted chicken meat (b). For use in carbohydrate and milk protein intolerance in infancy.

Complamex. Peripheral vasodilator: see **Xanthinol nicotinate.**

CVK (Compocillin VK). Antibiotic: see **Phenoxymethylpenicillin.**

Concavit. Vitamin mixture: see **Vitamin A, Thiamine, Riboflavine, Pyridoxine, Cyanocobalamin, Vitamin C, Nicotinamide, Pantothenic acid.**

Concordin.   Antidepressant: see **Protriptyline.**
Conotrane.   Topical anti-infective: see **Hydrargaphen.**
Conovid/Conovid E.   Oral contraceptives: see **Mestranol, Norethynodrel.**
Controvlar.   For menstrual irregularities: see **Ethinyloestradiol, Norethisterone.**
Coolspray.   Rubefacient spray containing fluorinated hydrocarbons.
Copholco.   Cough suppressant: see **Pholcodine.**
Co-Pyronil.   Anti-allergic: see **Pyrrobutamine, Methapyrilene, Cyclopentamine.**
Coramine (d).   C.N.S. stimulant: see **Nikethamide.**
Corangil (d).   Anti-angina: see **Glyceryl trinitrate, Diprophylline, Papaverine, Pentaerythritol tetranitrate.**
Cordex.   Anti-inflammatory/analgesic: see **Prednisolone, Acetylsalicylic acid.**
Cordilox.   Anti-anginal: see **Verapamil.**
Cordocel.   Anti-infective powder for skin, wounds: see **Alum, Zinc powder, Hexachlorophane.**
Corlan.   Corticosteroid pellet for aphthous ulcers: see **Hydrocortisone.**
Corsodyl.   Topical treatment for gingivitis: see **Chlorhexidine.**
Cor-Tar-Quin.   Topical corticosteroid/anti-infective: see **Hydrocortisone, Iodohydroxyquinoline.**
Cortelan.   Corticosteroid: see **Cortisone.**
Cortenema.   Corticosteroid enema for colitis: see **Hydrocortisone.**
Cortico-Gel.   Corticotrophic hormone injection: see **Corticotrophin.**
Cortistab.   Corticosteroid: see **Cortisone.**
Cortisyl.   Corticosteroid: see **Cortisone.**
Cortril.   Topical corticosteroid: see **Hydrocortisone.**
Cortrophin ZN.   Corticotrophic hormone injection: see **Corticotrophin.**
Cortrosyn depot.   Synthetic corticotrophic injection: see **Tetracosactrin.**
Cortucid.   Corticosteroid eye drops: see **Sulphacetamide, Hydrocortisone.**
Cosaldon.   Peripheral vasodilator: see **Pentifylline, Nicotinic acid.**
Cosmegen Lyovac.   Cytotoxic: see **Actinomycin D.**
Cosylan.   Cough suppressant: see **Dextromethorphan.**
Cotazym.   For use in pancreatic deficiency: see **Pancreatic enzymes.**
Covering Cream (b).   Concealing cream: see **Titanium dioxide.**
Covermark (b).   Concealing cream: see **Titanium dioxide.**
Crasnitin.   Cytotoxic: see **L-Asparaginase.**
Cremalgex.   Rubefacient: see **Methyl nicotinate, Glycol salicylate, Histamine.**
Cremalgin.   Rubefacient: see **Methyl nicotinate, Glycol salicylate, Histamine.**
Cremathurm R.   Rubefacient: see **Ethyl nicotinate, Methyl salicylate, Histamine.**
Cremomycin.   Antibacterial/antidiarrhoeal: see **Succinylsulphathiazole, Neomycin, Kaolin.**
Cremostrep.   Antibacterial/antidiarrhoeal: see **Succinylsulphathiazole, Streptomycin, Kaolin.**
Cremosuxidine.   Antibacterial/antidiarrhoeal: see **Succinylsulphathiazole, Kaolin.**
Crinagen (b) (d).   Shampoo for scalp dermatitis: see **Triclocarban, Salicylic acid.**

Crookes ACTH/CMC (d).   Corticotrophic hormone injection: see **Corticotrophin.**

Crystamycin.   Antibiotic: see **Benzylpenicillin, Streptomycin.**

Crystapen.   Antibiotic: see **Benzylpenicillin.**

Crystapen G.   Antibiotic: see **Benzylpenicillin.**

Crystapen V.   Antibiotic: see **Phenoxymethyl penicillin.**

Cuemid.   Lowers serum cholesterol and bile acids: see **Cholestyramine.**

Cuprimine.   Chelating agent for Wilson's disease, cystinuria: see **Penicillamine.**

Cutisan.   Topical anti-infective: see **Triclocarban.**

Cyclimorph (c).   Narcotic analgesic: see **Morphine, Cyclizine.**

Cyclogest.   Suppository for treatment of premenstrual symptoms: see **progesterone.**

Cyclomet.   Sedative/tranquilliser: see **Cyclobarbitone, Emetine.**

Cyclo-Progynova.   Sex hormones for menopausal symptoms: see **Oestradiol, Norgestrel.**

Cycloserine.   Antibiotic used in tuberculosis and in urinary tract infections: see **Cycloserine.**

Cyclospasmol.   Peripheral vasodilator: see **Cyclandelate.**

Cylert.   C.N.S. stimulant: see **Pemoline.**

Cymogran (b).   For phenylketonuria diet.

Cytacon.   For vitamin B_{12} deficiency: see **Cyanocobalamin.**

Cytamen.   For vitamin B_{12}-deficient anaemias: see **Cyanocobalamin.**

Cytosar.   Cytotoxic: see **Cytarabine.**

D

Dactil.   Gut antispasmodic: see **Piperidolate.**

Daktacort.   Topical anti-infective/corticosteroid: see **Micronazole, Hydrocortisone.**

Daktarin.   Topical antifungal for skin and nails: see **Miconazole.**

Dalacin C.   Antibiotic: see **Clindamycin.**

Dalmane.   Hypnotic: see **Flurazepam.**

Dalysep (d).   Antibacterial: see **Sulfametopyrazine.**

Daneral-SA.   Anti-allergic: see **Pheniramine.**

Danol.   Semisynthetic steroid used in endometriosis: see **Danazol.**

Dantrium.   Muscle relaxant: see **Dantrolene.**

Daonil.   Oral hypoglycaemic: see **Glibenclamide.**

Daranide.   Diuretic used in states of respiratory acidosis: see **Dichlorphenamide.**

Daraprim.   Antimalarial: see **Pyrimethamine.**

Daricon.   Gut antispasmodic: see **Oxyphencyclamine.**

Dartalan.   Tranquilliser/anti-emetic: see **Thiopropazate.**

Davenol.   Decongestant/cough suppressant: see **Carbinoxamine, Ephedrine, Pholcodeine.**

DDAVP.   Synthetic antidiuretic hormone: see **Desmopressin.**

Deanase. Enzyme injection for local inflammatory disorders: see **Desoxy-ribonuclease.**

Deanase D.C. Oral enzyme for inflammatory disorders: see **Chymotrypsin.**

Debendox. Antiemetic: see **Dicyclomine, Doxylamine, Pyridoxine.**

Decadron. Corticosteroid: see **Dexamethasone.**

Deca-Durabolin. Anabolic steroid: see **Nandrolone.**

Decaserpyl. Antihypertensive: see **Methoserpidine.**

Decaserpyl plus. Antihypertensive: see **Methoserpidine, Benzthiazide.**

Decaspray. Corticosteroid/antibiotic aerosol: see **Dexamethasone, Neomycin.**

Declinax. Antihypertensive: see **Debrisoquine.**

Decoderm. Corticosteroid cream: see **Fluprednylidene.**

Decorpa. Anti-obesity bulk agent: see **Guar gum.**

Decortisyl. Corticosteroid: see **Prednisone.**

Defencin CP. Cerebral and peripheral vasodilator: see **Isoxsuprine.**

Degranol. Cytotoxic: see **Mannomustine.**

Dehydrocholin. Laxative: see **Dehydrocholic acid.**

Delfen. Spermicidal contraceptive: see **Nonoxynol-9.**

Delimon. Analgesic: see **Morazone, Paracetamol, Salicylamide.**

Delta-Butazolidin. Anti-inflammatory/analgesic: see **Phenylbutazone, Prednisone.**

Delta-Cortef. Corticosteroid: see **Prednisolone.**

Deltacortone. Corticosteroid: see **Prednisone.**

Deltacortril. Corticosteroid tablets or injection: see **Prednisolone.**

Deltacortril-enteric. Corticosteroid. Enteric coating said to reduce gastric irritation: see **Prednisolone.**

Deltastab. Corticosteroid tablets or injection: see **Prednisolone.**

Demulen/Demulen 50. Oral contraceptive: see **Mestranol, Ethynodiol.**

Dendrid. Local treatment for herpetic eye infections: see **Idoxuridine.**

De-Nol. Antacid: see **Tri-potassium di-citrato bismuthate.**

Depamine. Chelating agent used in severe rheumatoid arthritis: see **Penicillamine.**

Depixol. Major tranquilliser: see **Flupenthixol.**

Depocillin. Antibiotic: see **Procaine penicillin.**

Depo-Medrone. Corticosteroid for intra-articular/intramuscular injection: see **Methylprednisolone.**

Depo-Provera. Progestogen, depo-injection used in threatened abortion, endometriosis: see **Medroxyprogesterone.**

Depostat. Progestogen, depo-injection for benign prostatic hypertrophy: see **Gestronol.**

Depronal S.A. Analgesic: see **Dextropropoxyphene.**

Depropanex. Deproteinated pancreatic extract with vasodilator and antispasmodic properties.

Dequadin. Antiseptic throat lozenges: see **Dequalinium.**

Derbac liquid. Topical treatment for head lice: see **Malathion.**

Derbac shampoo. Shampoo for head lice: see **Carbaryl.**

Dermacaine. Topical anaesthetic cream: see **Cinchocaine.**

Dermalex. Soothing antiseptic lotion for prevention of bed sores or urine rash: see **Squalane, Hexachlorophane, Allantoin.**

Dermamed. Topical anti-infective ointment: see **Bacitracin, Neomycin.**
Dermogesic. Topical treatment for relief of itching and minor skin irritation, e.g. sunburn: see **Calamine, Cresol, Benzocaine.** The last may cause hypersensitivity rashes.
Dermonistat. Topical antifungal: see **Miconazole.**
Dermoplast (d). Aerosol local anaesthetic for pain relief after perineal surgery: see **Benzocaine, Benzethonium, Hydroxyquinoline, Menthol.**
Dermovate. Topical steroid treatments for psoriasis: see **Clobetasol.**
Dermovate-NN. Topical corticosteroid/anti-infective: see **Clobetasol, Neomycin, Nystatin.**
Deseril. Antiserotonin: see **Methysergide.**
Desferal. Chelating agent: see **Desferrioxamine.**
Deteclo. Antibiotic: see **Chlortetracycline, Tetracycline, Demeclocycline.**
Dettol. Topical antiseptic: see **Chloroxylenol.**
Dexacortisyl. Corticosteroid: see **Dexamethasone.**
Dexamed (c). C.N.S. stimulant: see **Dexamphetamine.**
Dexamethasone acetate. Corticosteroid capsules for inhalation in chronic rhinitis: see **Dexamethasone.**
Dexa-Rhinaspray. Nasal spray for allergic or chronic rhinitis: see **Tramazoline, Dexamethasone, Neomycin.**
Dexedrine (c). C.N.S. stimulant: see **Dexamphetamine.**
Dexocodene (c). Analgesic: see **Acetylsalicylic acid, Aluminium hydroxide, Phenacetin.**
Dextraven 110/150. Plasma expanders: see **Dextrans.**
DF 118. Analgesic: see **Dihydrocodeine.** (c) injection but not tablets.
Diabetic Cough Mixture (d). Sugar-free cough linctus: see **Codeine, Pholcodeine, Butethamate.**
Diabinese. Oral hypoglycaemic: see **Chlorpropamide.**
Di-Adreson. Corticosteroid: see **Prednisone.**
Di-Adreson F. Corticosteroid: see **Prednisolone.**
Diamox. Diuretic: see **Acetazolamide.**
Dianabol. Anabolic steroid: see **Methandienone.**
Dibenyline. Alpha adrenoceptor blocker: see **Phenoxybenzamine.**
Dibotin. Oral hypoglycaemic: see **Phenformin.**
Diconal (c). Analgesic: see **Dipipanone.**
Dicynene. Haemostatic: see **Ethamsylate.**
Diganox Nativelle. Cardiac glycoside: see **Digoxin.**
Digitaline Nativelle. Caridac glycoside: see **Digitoxin.**
Dijex. Antacid: see **Aluminium hydroxide, Magnesium carbonate.**
Diloran. Antacid: see **Magnesium oxide, Aluminium hydroxide, magnesium carbonate, Dimethicone.**
Dilosyn. (d) Antihistamine: see **Methdilazine.**
Dimelor. Oral hypoglycaemic: see **Acetohexamide.**
Dimipressin. Antidepressant: see **Imipramine.**
Dimotane. Antihistamine: see **Brompheniramine.**
Dimotane LA. Sustained release formulation of Dimotane.
Dimotane Expectorant. Expectorant cough mixture: see **Guaiphenesin, Phenylephrine, Phenylpropanolamine, Brompheniramine.**

Dimotane expectorant DC (c).   Cough linctus/expectorant. Similar to Dimotane Expectorant plus **Dihydrocodeine.**

Dimotane with codeine.   Cough linctus similar to Dimotane Expectorant plus **Codeine.**

Dimotapp.   Antihistamine/decongestant for symptomatic treatment of common cold. Available as sustained-release and elixir formulations: see **Brompheniramine, Phenylephrine, Phenylpropanolamine.**

Dimyril.   Cough suppressant: see **Isoaminile citrate.**

Dindevan.   Anticoagulant: see **Phenindione.**

Dioctyl ear capsules.   Oil for removal of ear wax: see **Dioctyl sodium sulphosuccinate.**

Dioctyl-Medo.   Purgative: see **Dioctyl sodium sulphosuccinate.**

Dioderm.   Corticosteroid cream for inflammatory and allergic skin conditions: see **Hydrocortisone.**

Dioderm C.   Topical corticosteroid/anti-infective: see **Hydrocortisone, Clioquinol.**

Diodoquin (d).   Anti-infective: see **Di-iodohydroxyquinoline.**

Diovol.   Antacid: see **Aluminium hydroxide, Dimethicone, Magnesium hydroxide.**

Dipar.   Oral hypoglycaemic: see **Phenformin.**

Dipidolor (c).   Analgesic: see **Piritamide.**

Direma.   Diuretic: see **Hydrochlorothiazide.**

Disalcid.   Analgesic: see **Salsalate.**

Disipal.   Anticholinergic/antiparkinsonian: see **Orphenadrine.**

Di-Sipidin.   Hormone extract: see **Posterior pituitary extract.**

Distalgesic.   Analgesic: see **Paracetamol, Dextropropoxyphene.**

Distamine.   Chelating agent used in rheumatoid arthritis: see **Penicillamine.**

Distaquaine V-K.   Antibiotic: see **Phenoxymethylpenicillin.**

Dithrolan.   Topical treatment for psoriasis: see **Dithranol, Salicylic acid.**

Dixarit.   Migraine prophylactic: see **Clonidine.**

Dolagin (d).   Analgesic/tranquilliser: see **Codeine, Paracetamol, Butobarbitone.**

Dolosan.   Analgesic: see **Acetylsalicylic acid, Dextropropoxyphene.**

Doloxene.   Analgesic: see **Dextropropoxyphene.**

Doloxene compound.   Analgesic: see **Dextropropoxyphene, Acetylsalicylic acid, Caffeine.**

Doloxytal (d).   Analgesic/tranquilliser: see **Dextropropoxyphene, Amylobarbitone.**

Dome-Acne.   Topical treatment for acne: see **Sulphur, Salicylic acid, Resorcinol.**

Dome-Cort cream.   Corticosteroid cream: see **Hydrocortisone.**

Domical.   Antidepressant: see **Amitriptyline.**

Donnagel (d).   Antidiarrhoeal: see **Kaolin, Pectin, Hyoscyamine, Atropine, Hyoscine.**

Donnagel-PG (d).   Anti-diarrhoeal. As for Donnagel plus **Opium tincture.**

Donnagel with neomycin.   Antidiarrhoeal/antibacterial: see **Kaolin, Pectin, Hyoscyamine, Atropine sulphate, Hyoscine hydrobromide, Neomycin.**

Donnatal. Anticholinergic/sedative for gastro-intestinal disorders: see **Hyoscyamine, Atropine sulphate, Hyoscine hydrobromide, Phenobarbitone.**

Donnatal LA. Long-acting form of Donnatal.

Dopamet. Antihypertensive: see **Methyldopa.**

Dopram. Respiratory stimulant: see **Doxapram.**

Dorbanex. Purgative: see **Danthron, Poloxalene.**

Doriden. Hypnotic: see **Glutethimide.**

Dramamine. Anti-emetic: see **Dimenhydrinate.**

Drapolene. Topical anti-infective: see **Benzalkonium.**

Drenison. Topical corticosteroid: see **Flurandrenolone.**

Drinamyl (c). For anxiety/depression: see **Amphetamine, Amylobarbitone.**

Droleptan. Premedicant, major tranquilliser: see **Droperidol.**

Dromoran (c). Narcotic analgesic: see **Levorphanol.**

Droxalin. Antacid: see **Aluminium carbonate, Magnesium trisilicate.**

Dryptal. Diuretic: see **Frusemide.**

DTIC. Cytotoxic: see **Dacarbazine.**

Dulcodos. Laxative: see **Bisacodyl, Dioctyl sodium sulphosuccinate.**

Dulcolax. Laxative: see **Bisacodyl.**

Duo-Autohaler. Aerosol bronchodilator: see **Isoprenaline, Phenylephrine.**

Duofilm. Topical treatment for warts: see **Salicylic acid, Lactic acid.**

Duogastrone. Position-release preparation for duodenal ulcer: see **Carbenoxolone.**

Duphalac. Purgative: see **Lactulose.**

Duphaston. For dysmenorrhoea and endometriosis: see **Dydrogesterone.**

Durabolin. Anabolic steroid: see **Nandrolone.**

Duracreme. Spermicidal contraceptive: see **Nonoxynol.**

Duragel (d). Spermicidal contraceptive: see **Nonoxynol.**

Durenate. Antibacterial: see **Sulphamethoxydiazine.**

Duromine. Anti-obesity: see **Phentermine.**

Duromorph (c). Narcotic analgesic: see **Morphine.**

Durophet (c). Anti-obesity: see **Amphetamine.**

Durophet-M (c). Anti-obesity: see **Amphetamine, Methaqualone.**

Duvadilan. Peripheral vasodilator: see **Isoxsuprine.**

Dyazide. Diuretic combination: see **Triamterene, Hydrochlorothiazide.**

Dytac. Diuretic: see **Triamterene.**

Dytide. Diuretic combination: see **Triamterene, Benzthiazide.**

E

E 45 cream. Skin-protective, paraffin-based cream.

Econocil V-K. Antibiotic: see **Phenoxymethylpenicillin.**

Economycin. Antibiotic: see **Tetracycline.**

Edecrin. Diuretic: see **Ethacrynic acid.**

Edosol (b). Low-salt dietary powder.

Efcortelan. Topical corticosteroid: see **Hydrocortisone.**

Efcortelan- Near/eye drops (d). Topical anti-inflammatory/antibiotic: see **Hydrocortisone, Neomycin.**
Efcortesol inj. Intravenous corticosteroid: see **Hydrocortisone.**
Effico. 'Tonic': see **Thiamine, Nicotinamide, Strychnine, Caffeine.**
Elamol. Antidepressant: see **Tofenacin.**
Elestol. Anti-inflammatory/analgesic: see **Chloroquine, Prednisone, Acetylsalicylic acid.**
Elixir Gabail. Sedative: see **Valerian, Bromide, Chloral Hydrate.**
Eltroxin. Thyroid hormone: see **Thyroxine.**
Eludril. Antiseptic solution/aerosol for oral infections: see **Chlorhexidine, Amethocaine.**
Emeside. Anticonvulsant: see **Ethosuximide.**
Emetrol. Anti-emetic: see **Laevulose, Phosphoric acid.**
Emko. Spermicidal contraceptive: see **Benzethonium, Nonoxynol.**
Emprazil. Antipyretic/decongestant: see **Pseudoephedrine, Acetylsalicylic acid, Caffeine.**
Enavid. For dysmenorrhoea and endometriosis: see **Mestranol, Norethynodrel.**
Endoxana. Cytotoxic: see **Cyclophosphamide.**
Endrine. Nasal decongestant drops: see **Ephedrine.**
Enduron. Diuretic: see **Methylclothiazide.**
Enduronyl. Antihypertensive: see **Methylclothiazine, Deserpidine.**
Englate. Bronchodilator syrup: see **Theophylline.**
Entair. Bronchodilator: see **Theophylline.**
Enterfram. Antibiotic/antidiarrhoeal: see **Framycetin, Kaolin.**
Enteromide. Antibacterial/antidiarrhoeal: see **Calcium sulphaloxate.**
Entrosalyl. Anti-inflammatory/analgesic: see **Sodium salicylate.**
Envacar. Antihypertensive: see **Guanoxan.**
Enzypan. For use in pancreatic deficiency: see **Pepsin, Pancreatin.**
Epanutin. Anticonvulsant: see **Phenytoin.**
Ephynal. Tocopheryl: see **Vitamin E.**
Epilim. Anticonvulsant: see **Sodium valproate.**
Epodyl. Cytotoxic: see **Ethoglucid.**
Epontol. Anaesthetic: see **Propanidid.**
Eppy. Eye drops for glaucoma: see **Adrenaline.**
Epsikapron. Haemostatic: see **Aminocaproic acid.**
Equagesic. Anti-inflammatory/analgesic: see **Ethoheptazine, Meprobamate, Acetylsalicylic acid, Calcium carbonate.**
Equanil. Sedative/tranquilliser: see **Meprobamate.**
Equipose. Sedative/tranquilliser: see **Hydroxyzine.**
Equivert. Anti-emetic, antivertigo: see **Buclizine, Nicotinic acid.**
Eraldin (d). For cardiac dysrrhythmias: see **Practolol.**
Erycen. Antibiotic: see **Erythromycin.**
Erythrocin. Antibiotic: see **Erythromycin.**
Erythromid. Antibiotic: see **Erythromycin.**
Erythroped. Antibiotic: see **Erythromycin.**
Esbatal. Antihypertensive: see **Bethanidine.**
Esidrex. Diuretic: see **Hydrochlorothiazide.**

Esidrex-K.   As Esidrex plus **Potassium chloride** supplement.
Eskacef.   Antibiotic: see **Cephradine.**
Eskamel.   Cream for acne: see **Resorcinol, Sulphur.**
Eskornade.   Anticholinergic/sympathomimetic/antihistimine mixture for symptomatic treatment of common cold: see **Isopropamide, Phenylpropanolamine, Diphenylpyraline.**
Esoderm (b).   Shampoo for head lice: see **Gamma-benzene hexachloride, Dicophane.**
Estrovis.   For suppression of lactation: see **Quinestrol.**
Ethnine (d).   Cough suppressant: see **Pholcodeine.**
Ethulose.   High-calorie intravenous food supplement: see **Laevulose.**
Etophylate.   Bronchodilator: see **Acepifylline.**
Etophylate P.P. (d).   Bronchodilator: see **Acepifylline, Papaverine, Phenobarbitone.**
Eudemine   Antihypertensive/hyperglycaemic: see **Diazoxide.**
Euglucon.   Oral hypoglycaemic: see **Glibenclamide.**
Eugynon 30 and 50.   Oral contraceptive: see **Ethinyloestradiol, Norgestrel.**
Eumydrin.   Anticholinergic: see **Atropine methonitrate.**
Eurax.   Topical antipruritic: see **Crotamiton.**
Eurax-hydrocortisone.   As Eurax plus **Hydrocortisone.**
Eutonyl.   Antihypertensive: see **Pargyline.**
Evadyne.   Antidepressant: see **Butriptyline.**
Evidorm.   Hypnotic: see **Hexobarbitone, Cyclobarbitone.**
Exolan.   Topical treatments for psoriasis: see **Dithranol.**
Expansyl Spansule.   Slow-release bronchodilator: see **Ephedrine, Diphenylpyraline, Trifluoperazine.**
Expulin.   Cough linctus: see **Pholcodine, Ephedrine, Chlorpheniramine, Menthol.**
Extil compound linctus.   Cough linctus: see **Noscapine, Pseudoephedrine, Carbinoxamine.**
Exyphen.   Cough linctus: see **Brompheniramine, Guaiphenesin, Phenylephrine, Phenylpropanolamine.**

F

Fabahistin.   Antihistamine: see **Mebhydrolin.**
Falcodyl.   Cough linctus: see **Pholcodine, Ephedrine.**
Faringets.   Local anaesthetic lozenges for sore throats: see **Myristylbenzalkonium.**
Fazadon.   Skeletal muscle relaxant: see **Fazadinium.**
Fe-Cap.   Haematinic: see **Ferrous glycine sulphate.**
Fe-Cap C.   As Fe-Cap plus **Vitamin C.**
Fe-Cap folic.   As Fe-Cap plus **Folic acid.**
Fefol Spansule.   Slow-release haematinic: see **Ferrous sulphate, Folic acid.**
Fefol-Vit Spansule.   As Fefol Spansule plus vitamins: see **Aneurine, Riboflavine, Pyridoxine, Nicotinic acid, Pantothenic acid.**

Fel

Felsol. Bronchodilator: see **Phenazone, Caffeine, Grindelia, Mistletoe.**
Femergin. Vasoconstrictor for migraine: see **Ergotamine.**
Femerital. Analgesic for dysmenorrhoea: see **Ambucetamide, Paracetamol.**
Femulen. Oral contraceptive: see **Ethynodiol.**
Fenobelladine. Anticholinergic/sedative for peptic ulcers: see **Pheno-barbitone, Belladonna extract.**
Fenopron. Non-steroid anti-inflammatory/analgesic: see **Fenoprofen.**
Fenostil retard. Antihistamine: see **Dimethindene.**
Fenox. Nasal drops/spray for allergic rhinitis: see **Phenylephrine, Chlorbutol.**
Fentazin. Tranquilliser/anti-emetic: see **Perphenazine.**
Feospan Spansule. Slow-release haematinic: see **Ferrous sulphate.**
Ferfolic M. Haematinic for prevention of anaemia of pregnancy: see **Ferrous gluconate, Folic acid, Ascorbic acid.**
Ferfolic SV. Haematinic for folic acid-deficient anaemias. Similar contents to Ferfolic M but higher **Folic acid** content.
Fergluvite. Haematinic/vitamins. Similar to Ferfolic but no **Folic acid.**
Fergon. Haematinic: see **Ferrous gluconate.**
Ferlucon elixir (d). Haematinic/vitamin: see **Ferrous gluconate, Aneurine.**
Ferraplex B. Haematinic/vitamins: see **Ferrous sulphate, Vitamin C, Aneurine, Riboflavine, Nicotinic acid.**
Ferrocap. Slow-release haematinic: see **Ferrous fumarate, Thiamine.**
Ferrocap F 350. Slow-release haematinic: see **Ferrous fumarate, Folic acid.**
Ferrograd C. Slow-release haematinic/vitamin: see **Ferrous sulphate, Vitamin C.**
Ferrograd folic. Slow-release haematinic: see **Ferrous sulphate, Folic acid.** acid.
Ferro-Gradumet. Slow-release haematinic: see **Ferrous sulphate.**
Ferromyn. Haematinic: see **Ferrous succinate.**
Ferromyn B. Haematinic/vitamins: see **Ferrous succinate, Aneurine, Riboflavine, Nicotinic acid.**
Ferromyn S. Haematinic: see **Ferrous succinate, Succinic acid.**
Ferromyn S folic. Haematinic: see **Ferrous succinate, Succinic acid, Folic**
Fersaday. Slow-release haematinic: see **Ferrous fumarate.**
Fersamal. Haematinic: see **Ferrous fumarate.**
Fesovit Spansule. Slow-release haematinic/vitamins: see **Ferrous sulphate, Vitamin C, Aneurine, Riboflavine, Pyridoxine, Nicotinic acid, Pantothenic acid.**
Feximac. Anti-inflammatory cream for eczema or allergic dermatitis: see **Bufexamac.**
Ficoid 2/Ficoid 5. Corticosteroid cream/ointment of differing concentration for eczema and other skin conditions: see **Fluocortolone.**
Ficoid 2-plus/Ficoid 5-plus. As Ficoid 2/Ficoid 5 plus anti-infective: see **Clemizole, Hexachlorophenate.**
Filair. Sympthomimetic bronchodilator: see **Terbutaline.**
Filon (c). Anti-obesity: see **Phenbutrazate, Phenmetrazine.**
Finalgon. Rubefacient: see **Nonylic acid, Butoxyethyl nicotinate.**
Flagyl. Antibacterial: see **Metronidazole.**

Flagyl Compak.   Anti-infective: see **Metronidazole, Nystatin.**
Flamazine.   Anti-infective cream for burns: see **Silver sulphadiazine.**
Flar.   Compound vitamins for prevention of gastric side effects due to anti-biotics: see **Thiamine, Riboflavine, Pyroxidine, Cyanocobalamin, Nicotin-amide, Sodium panothenate, Inositol nicotinate, Folic acid, Liver extract.**
Flavelix.   Decongestant: see **Mepyramine, Ephedrine, Ammonium chloride, Sodium citrate.**
Flaxedil.   Muscle relaxant: see **Gallamine.**
Fletcher's Phosphate Enema.   Purgative: see **Sodium phosphate, Sodium acid phosphate.**
Flexazone.   Non-steroid anti-inflammatory/analgesic: see **Phenylbutazone.**
Flexical (b).   High-calorie food with vitamins and minerals.
Floraquin.   Anti-infective pessaries for vaginal infections: see **Di-iodohydroxyquinoline, Boric acid, Phosphoric acid.**
Florinef.   Adrenocorticosteroid: see **Fludrocortisone.**
Floxapen.   Antibiotic: see **Flucloxacillin.**
Fluanxol.   Antidepressant/tranquilliser: see **Flupenthixol.**
Fluothane.   Anaesthetic gas: see **Halothane.**
Fluviron.   Influenze vaccine for immunization.
Folex-350.   Haematinic: see **Ferrous fumarate, Folic acid.**
Folicin.   Haematinic for prevention of anaemia in pregnancy: see **Ferrous sulphate, Folic acid, Copper sulphate, Manganese sulphate.**
Folvron.   Haematinic: see **Folic acid, Ferrous sulphate.**
Forceval.   Multivitamin and mineral supplement.
Forceval protein (b).   Protein, vitamin and mineral supplements for low-sodium/low-fat diets.
Fortagesic.   Analgesic: see **Pentazocine, Paracetamol.**
Fortior.   Haematinic/vitamins: see **Ferrous sulphate, Aneurine, Riboflavine, Nicotinic acid, Vitamin C, Copper sulphate, Manganese citrate.**
Fortral.   Analgesic: see **Pentazocine.**
Fosfor.   'Tonic': see **Phosphorylcolamine.**
Framycort.   Topical anti-infective/corticosteroid: see **Framycetin/Hydro-cortisone.**
Framygen cream.   Topical anti-infective for ear, eye, skin: see **Framycetin.**
Framyspray.   Anti-infective aerosol for wounds: see **Neomycin, Poly-mixin B, Bacitracin.**
Franol.   Bronchodilator/sedative: see **Ephedrine, Theophylline, Pheno-barbitone.**
Franol expectorant.   Bronchodilator/sedative/expectorant. As Franol plus **Guaiphenesin.**
Franol plus.   As Franol plus **Thenyldiamine.**
Frusid.   Diuretic: see **Frusemide.**
Fucidin.   Antibiotic: see **Fusidic acid.**
Fucidin H oint. and gel.   Topical anti-infective/corticosteroid: see **Fusidic acid.**
Fucidin Intertulle.   Antibiotic gauze dressing: see **Fusidic acid.**
Fulcin 125 and 500.   Antifungal: see **Griseofulvin.**
Fungilin.   Antifungal: see **Amphotericin.**

Fun

Fungizone intravenous.   Antifungal injection: see **Amphotericin.**
Furacin.   Anti-infective: see **Nitrofurazone.**
Furadantin.   Urinary anti-infective: see **Nitrofurantoin.**
Furan.   Urinary anti-infective: see **Nitrofurantoin.**
Furoxone.   Antibacterial/antidiarrhoeal: see **Furazolidone.**
Fybogel.   Laxative for diverticular disease, constipation: see **Ispaghula husk.**
Fybranta.   Laxative for diverticular disease, constipation: see **Bran.**

G

Galactomin Preps. (b).   Low-lactose dietary supplement.
Galenomycin.   Antibiotic: see **Oxytetracycline.**
Galfer and Galfer F.A.   For iron- and folic acid-deficiency anaemias: see **Ferrous fumarate, Folic acid.**
Gantanol (d).   Antibacterial: see **Sulphamethoxazole.**
Gantrisin.   Antibacterial: see **Sulphafurazole.**
Garoin.   Anticonvulsant: see **Phenytoin, Phenobarbitone.**
Gastalar.   Antacid: see **Aluminium hydroxide, Magnesium carbonate.**
Gastrils.   Antacid: see **Aluminium hydroxide, Magnesium carbonate.**
Gastrocote.   Antacid: see **Aluminium hydroxide, Magnesium trisilicate, Sodium bicarbonate.**
Gastrovite.   For iron and calcium deficiency: see **Ferrous sulphate, Vitamin C, Calciferol, Calcium gluconate.**
Gatinar.   Purgative: see **Lactulose.**
Gaviscon.   Antacid: see **Magnesium trisilicate, Aluminium hydroxide, Sodium bicarbonate.**
Gefarnil.   For peptic ulcers: see **Gefarnate.**
Gelofusine.   Plasma expander: see **Gelatin, Sodium chloride.**
Gelusil.   Antacid: see **Aluminium hydroxide, Magnesium trisilicate.**
Genexol.   Spermicidal pessary: see **Spermicides.**
Genisol (b).   Topical treatment for scalp seborrhoeic dermatitis: see **Coal tar.**
Genticin.   Antibiotic for topical and parenteral use: see **Gentamicin.**
Gentisone HC.   Anti-inflammation/anti-infective drops for outer ear: see **Gentamicin, Hydrocortisone.**
Gentran.   Plasma expander: see **Dextrans, Sodium chloride.**
Gentran 40/Gentran 70.   Plasma expanders: see **Dextrans.**
Gerisom.   Hypnotic: see **Paracetamol, Amylobarbitone, Chlormezanone.**
Gestanin.   For recurrent abortion: see **Allylestrenol.**
Gestyl.   Hormone: see **Gonadotrophin.**
Gevral.   Multivitamins and minerals for general deficiencies.
Gina (d).   Anti-angina: see **Propatylnitrate.**
Glibenese.   Oral hypoglycaemic: see **Glipizide.**
Glucophage.   Oral hypoglycaemic: see **Metformin.**
Glutarol.   Topical treatment for warts: see **Glutaraldehyde.**
Glutenex (b).   **Gluten**-free, milk-free biscuits for coeliac disease.
Gluten Free Biscuits (b).   **Gluten**-free dietary aid for coeliac disease.

136

Glutril.   Oral hypoglycaemic: see **Glibornuride.**
Glycinal.   Antacid: see **Aluminium glycinate, Magnesium trisilicate.**
Glykola.   'Tonic': see **Caffeine, Glycerophosphate.**
Gonadotraphon.   Hormone: see **Gonadotrophin.**
Gondafon.   Oral hypoglycaemic: see **Glymidine.**
GPV.   Antibiotic: see **Phenoxymethylpenicillin.**
Graneodin.   Antibiotic: see **Neomycin, Gramicidin.**
Gravol.   Anti-emetic: see **Dimenhydrinate.**
Gregoderm.   Topical anti-infective/corticosteroid: see **Neomycin, Nystatin, Polymyxin B, Hydrocortisone.**
Grisovin.   Antifungal: see **Griseofulvin.**
Guanimycin susp. forte.   Antibacterial/antidiarrhoeal: see **Dihydrostreptomycin, Sulphaguanidine, Kaolin.**
Guanor.   Cough suppressant: see **Diphenhydramine, Ammonium chloride menthol.**
Gynaflex.   Local application for vaginal infections: see **Noxytiolin, Lignocaine.**
Gyno-Daktarin.   Local application for vaginal and penile fungal infections: see **Miconazole.**
Gynovlar 21.   Oral contraceptive: see **Ethinyloestradiol, Norethisterone.**

H

H.11.   Polypeptide extract of male urine used in otherwise untreatable carcinoma.
Haelan.   Topical corticosteroid: see **Flurandrenolone.**
Haelan-C.   Topical corticosteroid/anti-infective: see **Flurandrenolone, Clioquinol.**
Haemaccel.   Plasma expander: see **Gelatin, Sodium chloride.**
Haemostop.   Haemostatic: see **Naftazone.**
Halcicomb.   Topical corticosteroid/anti-infective for skin lesions: see **Halcinonide, Neomycin, Nystatin.**
Halciderm.   Topical corticosteroid for skin disorders: see **Halcinonide.**
Haldol.   Tranquilliser: see **Haloperidol.**
Haldrate.   Corticosteroid: see **Paramethasone.**
Harmogen.   Female sex hormone for deficiency states: see **Oestrone.**
Harmonyl.   Antihypertensive: see **Deserpidine.**
Havapen.   Antibiotic: see **Penamecillin.**
Haynon.   Antihistamine/anti-allergic: see **Chlorpheniramine.**
Hayphryn.   Nasal decongestant spray: see **Phenylephrine, Thenyldiamine.**
Heminevrin.   Sedative/hypnotic: see **Chlormethiazole.**
Hemoplex.   Vitamin B mixture for deficiency state.
Hepacon B_{12}.   For vitamin B_{12} deficiency and pernicious anaemia: see **Cyanocobalamin.**
Hepacon-B forte.   For vitamin B_{12} and other deficiencies: see **Cyanocobalamin, Thiamine, Folic acid.**

Hepacon Liver injectable. For anaemia: see **Liver extracts, Hydroxocobalamin.**

Hepacon-Plex. For vitamin B deficiency: see **Cyanocobalamin, Thiamine, Riboflavine, Pyridoxine, Nicotinamide, Pantothenic acid.**

Hepacort. Plus cream, suppositories. Anticoagulant/corticosteroid for eczema and pruritis: see **Heparin, Hydrocortisone.**

Heparin retard. Anticoagulant formulated for intramuscular or deep subcutaneous injection: see **Heparin.**

Heptonal. Bronchodilator/sedative: see **Acepifylline, Heptaminol, Phenobarbitone.**

Herpid. Antiviral: see **Idoxuridine.**

Hexopal. Peripheral vasodilator: see **Inositol nicotinate.**

HF (2) (b). Histidine-free food for histidinaemia.

Hibiscrub. Antiseptic cleansing solution for pre-operative preparation of hands: see **Chlorhexidine.**

Hibitane cream. Antiseptic for prevention of skin infections: see **Chlorhexidine.**

Hibitane lozenges. Antiseptic/local anaesthetic for infected sore throats: see **Chlorhexidine, Benzocaine.**

Himaizol. Low-cholesterol food for hypercholesterolaemia.

Hiprex. Urinary anti-infective: see **Hexamine.**

Hirudoid. Anticoagulant cream for bruising associated with superficial thrombophlebitis or trauma: see **Heparin.**

Histadyl E.C. Cough suppressant/decongestant elixir: see **Codeine, Ephedrine, Methapyrilene, Ammonium chloride.**

Histalix. Decongestant: see **Diphenhydramine, Ammonium chloride, Sodium citrate.**

Histalog. Histamine analogue for tests of gastric acid secretion: see **Betazole.**

Histofax. Cream for minor skin irritations, e.g. sunburn: see **Calamine, Chlorcyclizine.**

Histryl. Antihistamine: see **Diphenylpyraline.**

Honvan. Synthetic sex hormone: see **Stilboestrol.**

Hormofemin compound. Synthetic sex hormone/sedative for menopausal symptoms: see **Dienoestrol, Phenobarbitone, Bromvaletone, Theobromide, Calcium lactate.**

Hormofemin cream. Synthetic sex hormone for pruritis, acne: see **Dienoestrol.**

Hormonin. Sex hormones for menopausal symptoms: see **Oestriol, Oestrone, Oestradiol.**

HRF. For diagnostic use in delayed sexual development and failure of pituitary gland function: see **Gonadorelin.**

Humotet. Purified human antitetanus immunoglobulin for passive immunisation.

Hyalase. Enzyme injection for addition to intramuscular/subcutaneous injection to aid absorption and decrease pain: see **Hyaluronidase.**

Hycal. High-calorie (carbohydrate), protein-free liquid for use in protein-free, low-electrolyte diet.

Hycozon. Corticosteroid cream for eczema and other skin conditions: see **Hydrocortisone, Urea.**
Hydergine. For impaired mental function in the elderly: see **Dihydroergocornine, Dihydroergocristine, Dihydroergokryptine.**
Hydrea. Cytotoxic: see **Hydroxyurea.**
Hydrenox. Diuretic: see **Hydroflumethiazide.**
Hydrocortistab. Corticosteroid for parenteral or topical use: see **Hydrocortisone.**
Hydrocortisyl. Corticosteroid cream for eczema and other skin conditions: see **Hydrocortisone.**
Hydrocortone. Corticosteroid for systemic or topical use: see **Hydrocortisone.**
Hydroderm. Corticosteroid/antibacterial cream for eczema: see **Hydrocortisone, Neomycin, Bacitracin.**
Hydromet. Antihypertensive/diuretic: see **Methyldopa,Hydrochlorothiazide.**
Hydromycin-D. Topical corticosteroid/antibacterial: see **Prednisolone, Neomycin.**
Hydrosaluric. Diuretic: see **Hydrochlorothiazide.**
Hydrosaluric-K. As Hydrosaluric plus **Potassium chloride** supplement.
Hygroton. Diuretic: see **Chlorthalidone.**
Hygroton K. As Hygroton plus **Potassium chloride** supplement.
Hypertane. Antihypertensive: see **Rauwolfia alkaloids.**
Hypertane compound. Antihypertensive/diuretic: see **Rauwolfia, Ethiazide, Potassium chloride.**
Hypertensan. Antihypertensive: see **Rauwolfia alkaloids.**
Hypertensin. Vasoconstrictor for shock: see **Angiotensin.**
Hypon. Analgesic/antipyretic: see **Acetylsalicylic acid, Caffeine, Codeine, Phenolpthalein.**
Hypovase. Antihypertensive: see **Prazocin.**

I

Iberol. Haematinic/vitamins: see **Ferrous sulphate, Vitamin C, Group B Vitamins, Liver extract.**
Icipen. Antibiotic: see **Phenoxymethylpenicillin.**
Icthaband. Topical treatment for eczema: see **Zinc ichthammol.**
Iliadin-mini. Nasal decongestant: see **Oxymetazoline.**
Ilonium. Protective, carminative ointment: see **Colophony, Turpentine oil, Camphene.**
Ilosone. Antibiotic: see **Erythromycin.**
Ilotycin. Antibiotic: see **Erythromycin.**
Imbrilon. Non-steroid anti-inflammatory/analgesic: see **Indomethacin.**
Imferon. Haematinic: see **Iron dextran injection.**
Imodium. Antidiarrhoeal: see **Loperamide.**
Imperacin. Antibiotic: see **Oxytetracycline.**

Imuran. Cytotoxic: see **Azathioprine.**

Inapasade. Antituberculous antibacterials: see **Isoniazid, *para*-Aminosalicylic acid.**

Inderal. Beta adrenoceptor blocker: see **Propranolol.**

Indocid. Non-steroid anti-inflammatory/analgesic: see **Indomethacin.**

Influvac. Inactivated influenza virus vaccine.

Inolaxine. Purgative: see **Sterculia.**

Insidon. Anxiolytic/antidepressant: see **Opipramol.**

Insomnol. Hypnotic: see **Methylpentynol.**

Intal. For asthma: see **Sodium cromoglycate.**

Intal compound. As Intal plus **Isoprenaline.**

Integrin. Anxiolytic: see **Oxypertine.**

Interacton. Anti-allergic enzyme preparation.

Intralgin. Rubefacient/local anaesthetic for muscle pain: see **Salicylamide, Benzocaine.**

Intralipid. High-energy source (fats) for intravenous feeding.

Intraval sodium. Short-acting intravenous barbiturate hypnotic for induction of anaesthesia: see **Thiopentone sodium.**

Intropin. Intravenous infusion for treatment of shock: see **Dopamine.**

Inversine. Antihypertensive: see **Mecamylamine.**

Ionamin. Anti-obesity: see **Phentermine.**

Irofol C. Haematinic for prevention and treatment of anaemia of pregnancy: see **Ferrous sulphate, Folic acid, Vitamin C.**

Ironorm. Haematinic: see **Iron dextran injection.**

Ismelin. Antihypertensive: see **Guanethidine.**

Ismelin Navidrex K. Antihypertensive. As Ismelin plus **Cyclopenthiazide.**

Iso-Autohaler. Aerosol bronchodilator: see **Isoprenaline.**

Iso-Bronchisan. Aerosol bronchodilator: see **Isoprenaline, Ephedrine, Theophylline.**

Isogel. Purgative: see **Ispaghula.**

Isopto alkaline. Lubricant ('artificial tears') for dry eyes: see **Hypromellose.**

Isopto atropine. Long-acting mydriatic eye drops for cycloplegic refraction and uveitis: see **Atropine sulphate, Methylcellulose.**

Isopto carbachol. Miotic eye drops for glaucoma: see **Carbacol, Methylcellulose.**

Isopto carpine. Miotic eye drops for glaucoma: see **Pilocarpine, Methylcellulose.**

Isopto epinal. Eye drops for glaucoma: see **Adrenaline.**

Isopto frin. Lubricant/decongestant for inflamed (but not infected) eye: see **Methylcellulose, Phenylephrine.**

Isopto plain. Lubricant ('artificial tears') for dry eyes: see **Hypromellose.**

Isordil. Anti-anginal: see **Isosorbide dinitrate.**

Isoxyl. Antituberculous antibacterial: see **Thiocarlide.**

Ivax. Antidiarrhoeal/antibacterial: see **Kaolin, Neomycin.**

J

Jadit. Topical antifungal: see **Buclosamide, Salicylic acid.**
Jadit-H. Topical antifungal/anti-inflammatory: see **Buclosamide, Salicylic acid, Hydrocortisone.**
Jectofer. For iron-deficiency anaemia: see **Iron-sorbitol** injection.
Jectoral (d). For iron-deficiency anaemia: see **Ferrous glycine sulphate.**
Juvel. Vitamin mixture: see **Vitamin A, Calciferol, Thiamine, Riboflavine, Pyridoxine, Nicotinamide, Vitamin C.**

K

Kalium Durules (d). Potassium supplement: see **Potassium chloride.**
Kamillosan oint. Topical preparation for sore skin: see **Resorcinol.**
Kanfotrex. Local treatment for otitis externa: see **Kanamycin, Amphomycin, Hydrocortisone.**
Kannasyn. Antibiotic: see **Kanamycin.**
Kantrex. Antibiotic: see **Kanamycin.**
Kantrexil (d). Antidiarrhoeal/antibacterial: see **Kanamycin, Bismuth carbonate, Attapulgite.**
Kaodene. Antidiarrhoeal: see **Codeine, Kaolin.**
Kaomycin. Antidiarrhoeal/antibacterial: see **Neomycin, Kaolin.**
Kaopectate. Antidiarrhoeal: see **Kaolin.**
Kaovax (d). Antidiarrhoeal/antibacterial: see **Kaolin, Succinylsulphathiazole.**
Kartonium. Oral resin used in oedema, toxaemia of pregnancy: see **Ammonium polystyrine sulphonate.**
Karvol. Inhalation for nasal congestion: see **Menthol.**
Katorin. Potassium supplement: see **Potassium gluconate.**
Kay-Cee-L. Potassium supplement: see **Potassium chloride.**
K-Contin. Potassium supplement: see **Potassium chloride.**
Keflex. Antibiotic: see **Cephalexin.**
Keflen. Antibiotic: see **Cephalothin.**
Kefzol. Antibiotic: see **Cephazolin.**
Kelferon. For iron-deficiency anaemia: see **Ferrous glycine sulphate.**
Kelfizine W. Urinary anti-infective: see **Sulfametopyrazine.**
Kelfolate. For anaemia of pregnancy: see **Ferrous glycine sulphate, Folic acid.**
Kelocyanor. Antidote for cyanide poisoning: see **Cobalt tetracemate.**
Kemadrin. Antiparkinsonian: see **Procyclidine.**
Kemicetine. Antibiotic: see **Chloramphenicol.**
Kenalog. Corticosteroid injection for allergic conditions: see **Triamcinolone.**
Kerecid. Local treatment for herpetic eye infections: see **Idoxuridine.**
Keromask (b). Concealing cream: see **Titanium dioxide.**
Kest. Purgative: see **Magnesium sulphate, Phenolphthalein.**

6

Kethamed. C.N.S. stimulant: see **Pemoline.**

Ketovite. Vitamin mixture: see **Acetomenaphthone, Thiamine, Riboflavine, Pyridoxine, Nicotinamide, Pantethenate, Vitamin C, Tocopheryl, Folic acid.**

Kidnamin. Source of essential amino acids.

Kinidin Durules. Antidysrhythmic: see **Quinidine.**

Kloref. Effervescent potassium supplement: see **Potassium chloride.**

Koate. Clotting factor for intravenous injection in haemophilia: see **Factor VIII.**

Kolanticon. Antacid/antispasmodic: see **Aluminium hydroxide, Magnesium oxide, Dicyclomine.**

Kolantyl. Antacid/antispasmodic: see **Aluminium hydroxide, Magnesium oxide, Dicyclomine.**

Kolantyl-NV. Antacid/antispasmodic: see **Aluminium hydroxide, Magnesium hydroxide, Magnesium trisilicate, Dicyclomine.**

Konakion. For prothrombin deficiency: see **Phytomenadione.**

L

Labiton. 'Tonic': see **Thiamine, para-Aminobenzoic acid, Caffeine.**

Labophylline. Bronchodilator: see **Theophylline.**

Laboprin. Anti-inflammatory/analgesic: see **Acetylsalicylic acid.**

Labosept. Oral antiseptic: see **Dequalinium.**

Laevulfex. Intravenous energy source: see **Laevulose.**

Lamprene. Antileprosy: see **Clafazimine.**

Lanitop. For cardiac failure: see **Medigoxin.**

Lanoxin. For cardiac failure: see **Digoxin.**

Lanvis. Cytotoxic: see **Thioguanine.**

Largactil. Major tranquilliser/anti-emetic/antivertigo: see **Chlorpromazine.**

Larodopa. Antiparkinsonian: see **Laevodopa.**

Lasix. Diuretic: see **Frusemide.**

Lasonil. Topical treatment for soft tissue injury: see **Heparinoid, Hyaluronidase.**

Laxoberal. Purgative: see **Sodium picosulphate.**

Ledclair. For heavy metal poisoning: see **Sodium calcium edetate.**

Ledercort. Corticosteroid for topical or systemic use: see **Triamcinolone.**

Lederkyn. Antibacterial: see **Sulphamethoxypyridazine.**

Ledermycin. Antibiotic: see **Demeclocycline.**

Lederplex. Vitamin B mixture: see **Thiamine, Riboflavine, Pyridoxine, Cyanocobalamin, Pantothenic acid, Niacinamide.**

Lederspan. Corticosteroid injection: see **Triamcinolone.**

Lederstatin (d). Antibiotic/antifungal combination: see **Demeclocycline, Nystatin.**

Lenium (b). Shampoo for dandruff: see **Selenium sulphide.**

Lentizol. Antidepressant: see **Amitriptyline.**

Leo K. Slow-release potassium supplement: see **Potassium chloride.**

Lergoban.   Anti-allergic: see **Diphenylpyraline.**
Lethidrone.   Narcotic antagonist injection: see **Nalorphine.**
Leucovorin.   Source of folinic acid for deficiency; antagonises antifolate cytotoxics.
Leukeran.   Cytotoxic: see **Chlorambucil.**
Levius.   Anti-inflammatory/analgesic: see **Acetylsalicylic acid.**
Levophed.   Vasoconstrictor: see **Noradrenaline.**
Libraxin.   Anxiolytic antispasmodic: see **Chlordiazepoxide, Clidinium.**
Librium.   Anxiolytic: see **Chlordiazepoxide.**
Lidothesin.   Local anaesthetic: see **Lignocaine.**
Limbritol 10 and 5.   Antidepressant/anxiolytic: see **Amitriptyline, Chlordiazepoxide.**
Limclair.   Increases calcium excretion: see **Sodium edetate.**
Lincocin.   Antibiotic: see **Lincomycin.**
Linctifed.   Decongestant/mucolytic: see **Triprolidine, Pseudoephedrine, Codeine, Guaiphenesin.**
Lingraine.   Vasoconstrictor for migrane: see **Ergotamine.**
Lioresal.   Muscle relaxant: see **Baclofen.**
Lipiphysan (d).   High-energy, intravenous infusion for parenteral feeding.
Lipoflavonoid.   Multi-vitamin preparation for Ménière's disease.
Lipotriad.   Multiple vitamins for retinal degeneration.
Liprinal.   Lipid-lowering agent: see **Clofibrate.**
Lobak.   Analgesic: see **Paracetamol, Chlormezanone.**
Locabiotal.   Topical antibiotic for infections of upper respiratory tract: see **Fusafungine.**
Locan.   Topical anaesthetic cream: see **Amylocaine, Amethocaine, Cinchocaine.**
Locasol (b).   Low-calcium food substitute for calcium intolerance.
Locoid.   Topical corticosteroid for eczema and other skin conditions: see **Hydrocortisone.**
Locorten (d).   Topical corticosteroid for eczema and other skin conditions: see **Flumethasone.**
Locorten/Vioform.   Topical corticosteroid/anti-infective for skin or ears: see **Flumethasone, Clioquinol.**
Loestrin 20.   Oral contraceptive: see **Ethinyloestradiol, Norethisterone.**
Lofenalac (b).   Dietary substitute for phenylketonuria.
Lomodex 40 and 70.   Plasma expanders: see **Dextrans.**
Lomotil with neomycin.   Antidiarrhoeal/antibacterial: as Lomotil plus **Neomycin.**
Lomotil with neomycin.   Antidiarrhoeal/antibacterial: see **Neomycin.**
Lomupren (d).   Sympathomimetic bronchodilator powder for inhalation: see **Isoprenaline.**
Lomusol.   Anti-allergic nasal decongestant: see **Sodium chromoglycate.**
Lopresor.   Beta adrenoceptor blocker: see **Metoprolol.**
Lorexane.   Lotion for head lice: see **Gamma-benzene hexachloride.**
Lorfan.   Narcotic antagonist: see **Levallorphan.**
Lotussin.   Cough linctus/decongestant: see **Dextromethorphan, Guaiphenesin, Diphenhydramine, Ephedrine.**

LPLS (1) Formula (b).   Dietary substitute, low in protein, salt.
LPF (1) (b).   Dietary substitute, low in phenylalamine, tyrosine.
LPTM (2) (b).   Dietary substitute, low in phenylalamine, tyrosine, methionine.
Lucidril.   C.N.S. stimulant for senile confusion: see **Meclofenoxate.**
Ludiomil.   Antidepressant: see **Maprotiline.**
Luminal.   Anticonvulsant/hypnotic: see **Phenobarbitone.**
Lyndiol 2.5 (d).   Oral contraceptive: see **Mestranol, Lynoestrenol.**
Lynoral.   Sex hormone for menopausal disorders, amenorrhoea, carcinoma of prostate: see **Ethinyloestradiol.**
Lyophrin.   Topical treatment for open-angle glaucoma: see **Adrenaline.**
Lysivane.   Anticholinergic/antiparkinsonian: see **Ethopropazine.**

M

M & B 693.   Antibacterial: see **Sulphapyridine.**
M & B antiseptic cream.   Topical anti-infective: see **Propamidine.**
Maalox.   Antacid: see **Aluminium hydroxide, Magnesium hydroxide.**
Macrodantin.   Antibacterial: see **Nitrofurantoin.**
Macrodex.   Plasma expander: see **Dextrans.**
Madecassol.   Soothing ointment and powder for sore skin: see **Centella asiatica.**
Madopar.   Antiparkinsonian: see **Laevodopa, Benserazide.**
Madribon.   Antibacterial: see **Sulphadimethoxine.**
Magmilor (d).   Antibacterial, antiprotozoal: see **Nifuratel.**
Magnapen.   Antibiotic: see **Ampicillin, Flucloxacillin.**
Malatex.   Keratolytic cream/solution for removal of thickened, dry skin: see **Propylene glycol, Malic acid, Benzoic acid, Salicylic acid.**
Maloprim.   Antimalarials for prophylaxis: see **Dapsone, Pyrimethamine.**
Mandelamine.   Urinary anti-infective: see **Hexamine mandelate.**
Mandrax (c).   Hypnotic: see **Methaqualone, Diphenhydramine.**
Marboran.   Antiviral: see **Methisazone.**
Marcain.   Local anaesthetic injection: see **Bupivacaine.**
Marevan.   Oral anticoagulant: see **Warfarin.**
Marplan.   Antidepressant (monoamine oxidase inhibitor): see **Isocarboxazid.**
Marsilid.   Antidepressant (monoamine oxidase inhibitor): see **Iproniazid.**
Masse.   Soothing cream for nipples during lactation: see **Aminoacridine, Allantoin.**
Masteril.   Sex hormone for treatment of neoplasms of the breast: see **Drostanolone.**
Maxidex.   Corticosteroid eye drops for non-infective, inflammatory conditions: see **Dexamethasone.**
Maxitrol.   Corticosteroid/antibiotic eye drops/ointment: see **Dexamethasone, Neomycin, Polymixin.**
Maxolon.   Anti-emetic: see **Metoclopramide.**

MCT (1) and MCT Oil (b). Dietary substitutes containing triglycerides. For use in impaired fat absorption.

Mebryl Spansule (d). Slow-release antihistamine: see **Embramine**.

Medihaler-Duo. Bronchodilator aerosol: see **Isoprenaline, Phenylephrine**.

Medihaler-Epi. Bronchodilator aerosol: see **Adrenaline**.

Medihaler-Ergotamine. Aerosol inhalation for migraine: see **Ergotamine**.

Medihaler-Iso/Medihaler Iso-Forte. Bronchodilator aerosol: see **Isoprenaline**.

Medocodene. Analgesic: see **Paracetamol, Codeine, Phenolphthalein**.

Medomet. Antihypertensive: see **Methyldopa**.

Medomin. Hypnotic: see **Heptabarbitone**.

Medro-cordex. Anti-inflammatory/analgesic: see **Methylprednisolone, Salicylic acid**.

Medrone. Corticosteroid: see **Methylprednisolone**.

Medrone acne lotion. Topical treatment for acne: see **Methylprednisolone, Sulphur, Aluminium Chlorohydrate**.

Medrone Medules. Slow-release corticosteroid. Claimed to reduce gastric side effects: see **Methylprednisolone**.

Medrone Veriderm. Topical corticosteroid/antibacterial: see **Methylprednisolone, Neomycin**.

Megaclor. Antibiotic: see **Clomocycline**.

Megimide. Respiratory stimulant: see **Bemegride**.

Melitase. Oral hypoglycaemic: see **Chlorpropamide**.

Melleril. Tranquilliser: see **Thioridazine**.

Meltrol. Oral hypoglycaemic: see **Phenformin**.

Menopax. Sex hormone/sedative for menopausal disorders: see **Ethinyloestradiol, Carbromal, Bromvaletone**.

Menopax forte. As Menopax plus **Methyltestosterone, Mephenesin**.

Menophase. Sex hormones for menopausal disorders: see **Mestranol, Norethisterone**.

Mepavlon (d). Anxiolytic: see **Meprobamate**.

Mepilin. Sex hormones for menopausal disorders or suppression of lactation: see **Ethinyloestradiol, Methyltestosterone**.

Merbentyl. Anticholinergic for gastro-intestinal colic: see **Dicyclomine**.

Merbentyl with phenobarbitone (d). As for Merbentyl plus **Phenobarbitone** as sedative.

Merocet. Antiseptic mouthwash or lozenges: see **Cetylpyridinium**.

Merthiolate. Antiseptic for pre-operative skin preparation or wound infections: see **Thiomersal**.

Mesontoin. Anticonvulsant: see **Methoin**.

Mestinon. Anticholinesterase: see **Pyrdiostigmine**.

Metabolic Mineral Mixture (b). Dietary mineral supplement.

Metamucil. Purgative: see **Psyllium**.

Metanium. Soothing, protective ointment and powder for nappy rash and other macerated skin conditions: see **Titanium dioxide**.

Metatone. 'Tonic': see **Thiamine, Glycerophosphate**.

Metenix S. Diuretic: see **Metolazone**.

Methisul. Antibacterial: see **Sulphamethizole**.

Metiguanide (d).   Oral antidiabetic: see **Metformin.**
Metilar.   Corticosteroid: see **Paramethasone.**
Metopirone.   Used in test of pituitary function: see **Metyrapone.**
Metosyn.   Topical corticosteroid for skin disorders: see **Fluocinonide.**
Metrulen.   For endometriosis, functional uteric bleeding: see **Ethynodiol, Mestranol.**
Metrulen 50.   For endometriosis, functional uterine bleeding: see **Ethinyloestradiol, Ethynodiol.**
Mevilin-L.   Live attenuated measles virus vaccine.
Mexitil.   Antidysrhythmic: see **Mexiletine.**
Micoren.   C.N.S. stimultant: see **Cropropamide, Crotethamide.**
Micralax.   Purgative enema: see **Sodium citrate, Sodium alkyl sulphoacetate, Sorbic acid.**
Microgynon 30.   Oral contraceptive: see **Ethinyloestradiol, Norgestrel.**
Micronor.   Oral contraceptive: see **Norethisterone.**
Midamor.   Diuretic: see **Amiloride.**
Midicel.   Antibacterial: see **Sulphamethoxypyridazine.**
Midrid.   For headache, migraine: see **Isometheptene, Paracetamol, Dichloralphenazone.**
Migen.   Extract of house dust mite for allergy densitisation.
Migraleve.   For migraine: see **Buclizine, Paracetamol, Codeine, Dioctyl sodium sulphosuccinate.**
Migril.   Vasoconstrictor for migraine: see **Ergotamine, Cyclizine, Caffeine.**
Millophyline.   Cardiac stimulant/bronchodilator: see **Etamiphylline.**
Milonorm.   Sedative/tranquilliser: see **Meprobamate.**
Miltown.   Sedative/tranquilliser: see **Meprobamate.**
Minafen (b).   Dietary preparation for phenylketonuria.
Minamino.   Mixture of minerals and vitamins for deficiency states.
Minihep.   Subcutaneous injection for thrombo-embolic disease: see **Heparin.**
Minilyn.   Oral contraceptive: see **Ethinyloestradiol, Lynoestrenol.**
Minims.   Single dose ophthalmic preparations.
Minocin.   Antibiotic: see **Minocycline.**
Minodiab.   Oral hypoglycaemic: see **Glipizide.**
Minovlar.   Oral contraceptive: see **Ethinyloestradiol, Norethisterone.**
Mintezol.   Anthelmintic: see **Thiabendazole.**
Miochol.   Miotic eydrops: see **Acetylcholine.**
Mio-Pressin (d).   Antihypertensive: see **Rauwolfia, Protoveratrine, Phenoxybenzamine.**
Mixogen.   Sex hormones for menopausal symptoms, suppression of lactation: see **Ethinyloestradiol, Methyltestosterone.**
Modecate.   Long-acting tranquilliser injection: see **Fluphenazine.**
Moditen.   Tranquilliser: see **Fluphenazine.**
Moditen enanthate.   Long-acting tranquilliser injection: see **Fluphenazine.**
Moduretic.   Diuretic combination: see **Amiloride, Hydrochlorothiazide.**
Mogadon.   Hypnotic: see **Nitrazepam.**
Molcer.   Drops to soften ear wax: see **Dioctyl sodium sulphosuccinate.**
Molivate.   Topical corticosteroid for skin disorders: see **Clobetasone.**

Monistat. Topical antifungal: see **Miconazole.**
Monodral. Anticholinergic for reducing gastric motility and secretion: see **Penthienate.**
Monotheamin. Bronchodilator: see **Theophylline.**
Monophytol. Topical antifungal: see **Boric acid, Chlorbutol, Salicylic acid, Undecenoic acid.**
Moore's Teejel. For oral ulcers: see **Choline salicylate, Cetalkonium.**
Morhulin. Topical preparation for abrasions, skin ulcers: see **Zinc oxide, Cetrimide, Dakin's solution.**
Morsep. Topical preparation for nappy rash: see **Cetrimide, Dakin's solution.**
Motipress. Sedative/antidepressant: see **Fluphenazine, Nortriptyline.**
Motival. Sedative/antidepressant: see **Fluphenazine, Nortriptyline.**
Movelat. Rubefacient: see **Corticosteroid, Salicylic acid.**
M.S.U.D. Aid (b). Dietary aid for maple syrup urine disease.
Mucaine. Antacid for oesophageal pain: see **Oxethazaine, Aluminium hydroxide, Magnesium hydroxide.**
Mucodyne. Expectorant: see **Carbocisteine.**
Muflin. Cough suppressant: see **Dextromethorphan, Pheniramine.**
Multibionta. Intravenous vitamins: see **Vitamin A, Thiamine, Riboflavine, Nicotinamide, Pantothenic acid, Vitamin C, Pyridoxine, Tocopheryl.**
Multivite. Vitamin mixture: see **Vitamin A, Thiamine, Vitamin C, Calciferol.**
Mumpsvax. Live mumps virus vaccine.
Muripsin. Preparation of hydrochloric acid and **Pepsin** for deficient gastric secretion.
Myambutol. Antituberculosis: see **Ethambutol.**
Myanesin. For muscle spasm: see **Mephenesin.**
Mycardol. Anti-anginal: see **Pentaerythritol tetranitrate.**
Mycifradin. Antibiotic: see **Neomycin.**
Myciguent. Topical antibiotic: see **Neomycin.**
Mycil. Topical antifungal: see **Chlorphenesin.**
Mycivin. Antibiotic: see **Lincomycin.**
Mycolactine. Purgative: see **Aloes.**
Mycota. Topical antifungal: see **Undecenoic acid.**
Mydriacyl. Mydriatic/cycloplegic eye drops: see **Tropicamide.**
Mydrilate. Mydriatic/cycloplegic eye drops: see **Cyclopentolate.**
Myelobromol. Cytotoxic: see **Mitobronitol.**
Mylanta. Antacid: see **Magnesium hydroxide, Aluminium hydroxide.**
Myleran. Cytotoxic: see **Busulphan.**
Mynah. Antituberculosis: see **Ethambutol, Isoniazid.**
Myocrisin. Gold injection for rheumatoid arthritis: see **Aurothiomalate sodium.**
Myolgin. Analgesic: see **Paracetamol, Codeine, Caffeine, Acetomenaphthone.**
Myotonine chloride. Produces gut and bladder emptying: see **Bethanechol.**
Mysoline. Anticonvulsant: see **Primidone.**
Mysteclin. Antibiotic/antifungal: see **Tetracycline, Nystatin.**
Mytelase. For myasthenia gravis: see **Ambenonium.**

N

Nactisol (d).   Anticholinergic/sedative for peptic ulceration: see **Poldine, Secbutobarbitone.**

Nacton/Nacton forte.   Anticholinergic for peptic ulcers: see **Poldine.**

Naprosyn.   Non-steroid anti-inflammatory/analgesic: see **Naproxen.**

Napsalgesic.   Analgesic: see **Acetylsalicylic acid, Dextropropoxyphene.**

Narcan.   Narcotic antagonist: see **Naloxone.**

Nardil.   Antidepressant (monoamine oxidase inhibitor): see **Phenelzine.**

Narex.   Nasal decongestant: see **Phenylephrine, Chlorbutol.**

Narphen (c).   Analgesic: see **Phenazocine.**

Naseptin.   Antiseptic/antibacterial cream for topical use in nasal carriers of staphylococci: see **Chlorhexidine, Neomycin.**

Nasoflu (d).   Live attenuated influenza virus for immunisation.

Natirose.   Anti-anginal: see **Glyceryl trinitrate, Ethylmorphine, Hyoscyamine hydrobromide.**

Natisedine (d).   Antidysrhythmic/sedative: see **Quinidine, Phenobarbitone.**

Natulan.   Cytotoxic: see **Procarbazine.**

Navane.   Tranquilliser: see **Thiothixene.**

Navidrex.   Diuretic: see **Cyclopenthiazide.**

Navidrex-K.   As Navidrex plus **Potassium chloride** in slow-release wax core.

Naxogin/Naxogin 500.   Antiprotozoal: see **Nimorazole.**

Nebcin.   Antibiotic: see **Tobramycin.**

Nefranutrin (b).   Essential amino acids. Dietary supplement for chronic renal failure.

Nefrolan.   Diuretic: see **Chlorexolone.**

Negram.   Antibacterial: see **Nalidixic acid.**

Nembutal.   Hypnotic: see **Pentobarbitone.**

Neobacrin oint.   Topical anti-infective: see **Neomycin, Bacitracin.**

Neo-Cantil (d).   Anticholinergic/antibacterial. For infective diarrhoea: see **Mepenzolate, Neomycin.**

Neo-Cortef.   Topical corticosteroid/antibacterial ointment/lotion/drops: see **Hydrocortisone, Neomycin.**

Neo-Cytamen.   Vitamin: see **Hydroxocobalamin.**

Neogest.   Oral contraceptive: see **Norgestrel.**

Neo-Medrone.   Topical lotion for acne: see **Methylprednisolone, Neomycin, Sulphur, Aluminium chlorohydrate.**

Neo-Medrone Veriderm.   Corticosteroid ointment for inflammatory/allergic skin conditions: see **Methylprednisolone.**

Neo-Mercazole.   Antithyroid: see **Carbimazole.**

Neomin.   Antibiotic: see **Neomycin.**

Neo-Naclex.   Diuretic: see **Bendrofluazide.**

Neo-Naclex-K.   As Neo-Naclex plus **Potassium chloride** in slow-release matrix.

Neophryn.   Nasal decongestant: see **Phenylephrine.**

Neosporin.   Antibacterial eye drops: see **Polymyxin B, Neomycin, Gramicidin.**

Neo-Sulfazon. Antidiarrhoeal/antibacterial: see **Neomycin, Phthalysul-phathiazole, Kaolin, Pectin.**
Neovax. Antidiarrhoeal/antibacterial: see **Neomycin, Kaolin.**
Nephril. Diuretic: see **Polythiazide.**
Neptal. Diuretic for intramuscular injection: see **Mercuramide, Theophylline.**
Nerisone. Corticosteroid cream for skin conditions: see **Diflucortolone.**
Nestosyl. Topical ointment for painful, itchy skin conditions: see **Ben-zocaine, Butyl aminobenzoate, Resorcinol, Zinc oxide, Hexachlorophane.**
Nethaprin Dospan. Slow-release bronchodilator: see **Etafedrine, Bufylline, Doxylamine, Phenylephrine.**
Nethaprin expectorant. As Nethaprin minus phenylephrine but plus **Glyceryl guaiacolate** as expectorant.
Neulactil. Tranquilliser: see **Pericyazine.**
Neurodyne. Analgesic: see **Paracetamol, Codeine.**
Neuro-Phosphates. 'Tonic': see **Glycerophosphates, Strychnine.**
Neutradonna. Antacid/anticholinergic for peptic ulcers: see **Aluminium antacids, Belladonna alkaloids.**
Neutradonna Sed. As Neutradonna plus **Amylobarbitone.**
Neutraphylline. Bronchodilator: see **Diprophylline.**
Neutrolactis. Antacid: see **Aluminium hydroxide, Magnesium trisilicate, Calcium carbonate.**
Niamid. Antidepressant: see **Nialamide.**
Niferex. Haematinic: see **Polysaccharide-iron complex.**
Nilevar. Anabolic steroid: see **Norethandrolone.**
Nilstim. Anti-obesity bulk agent: see **Methylcellulose.**
Nitrocontin. Anti-anginal: see **Glyceryl trinitrate.**
Nivaquine. Antimalarial: see **Chloroquine.**
Nivembin. Anti-amoebic: see **Chloroquine, Di-iodohydroxyquinoline.**
Nivemycin. Antibiotic for oral and topical use: see **Neomycin.**
Nobrium. Anxiolytic: see **Medazepam.**
Noctec. Hypnotic: see **Chloral hydrate.**
Noludar. Hypnotic: see **Methyprylone.**
Nolvadex. For treatment of anovular infertility and breast cancer: see **Tamoxifen.**
Nomaze. Nasal decongestant: see **Ephedrine, Naphazoline.**
Noradran. Bronchodilator expectorant: see **Guaiphenesin, Diphenhydra-mine, Diprophylline, Ephedrine.**
Noratex. Cream for bed sores: see **Talc, Kaolin (light), Zinc Oxide, Cod Liver Oil.**
Norfer. Haematinic: see **Ferrous fumarate.**
Norflex. Slow-release muscle relaxant: see **Orphenadrine.**
Norgesic. Muscle relaxant/analgesic: see **Paracetamol, Orphenadrine.**
Norgotin. Decongestant/anaesthetic/anti-infective for inflammation of outer ear: see **Ephedrine, Amethocaine, Chlorhexidine.**
Noriday. Oral contraceptive: see **Norethisterone.**
Norinyl-1/Norinyl-1/28. Oral contraceptive: see **Mestranol, Norethisterone.**
Norinyl-2. As Norinyl-1 but larger dosage for control of heavy or irregular menstruation.

Norlestrin.   For menstrual disorders: see **Ethinyloestradiol, Norethisterone.**
Normacol.   Purgative: see **Sterculia, Frangula.**
Normacol X.   Purgative: see **Sterculia, Danthron.**
Normax.   Purgative: see **Dioctyl sodium sulphosuccinate, Danthron.**
Norpace.   Cardiac antidysrhthmic see: **Disopyramide.**
Norpramine.   Antidepressant: see **Imipramine.**
Norval.   Antidepressant: see **Mianserin.**
Noveril.   Antidepressant: see **Dibenzepin.**
Noxyflex S.   Anti-infective for installation in bladder or other body cavities: see **Noxytiolin.**
Noxyflex with amethocaine.   As Nolyflex plus **Amethocaine** aslo cal anaesthetic.
Nuelin.   Bronchodilator: see **Theophylline.**
Nulacin.   Antacid: see **Magnesium antacids.**
Numotac.   Bronchodilator: see **Isoetharine.**
Nupercainal.   Topical skin anaesthetic: see **Cinchocaine.**
Nupercaine.   Topical skin anaesthetic: see **Cinchocaine.**
Nutramigen (b).   Dietary aid for lactose intolerance, galactosaemia.
Nutregen (b).   Dietary aid for coeliac disease, **Gluten** sensitivity.
Nutrizym.   For use in pancreatic deficiency: see **Pancreatic enzymes.**
Nydrane.   Anticonvulsant: see **Beclamide.**
Nystadermal.   Topical antifungal anti-inflammatory: see **Nystatin, Triamcinolone.**
Nystaform-HC.   Topical anti-infective anti-inflammatory: see **Nystatin, Clioquinol, Hydrocortisone.**
Nystan.   Antifungal: see **Nystatin.**
Nystavescent.   Vaginal antifungal: see **Nystatin.**

O

Oblivon.   Minor tranquilliser, sedative: see **Methylpentynol.**
Ocusert.   Pilo-20/Pilo-40. Sustained release cholinergic drug for treatment of glaucoma by placing under the eyelid: see **Pilocarpine.**
Ocusol.   Antibacterial eye drops: see **Sulphacetamide.**
Oestradin.   Sex hormone/sedative for menopausal disorders: see **Ethinyloestradiol, Phenobarbitone, Sodium bromide.**
Omnopon (c).   Narcotic analgesic: see **Papaveretum.**
Onadox-118.   Analgesic: see **Acetylsalicylic acid, Dihydrocodeine.**
Oncovin.   Cytotoxic: see **Vincristine.**
Operidine (c).   Narcotic analgesic: see **Phenoperidine.**
Ophthaine.   Topical ophthalmic anaesthetic: see **Proxymetacaine.**
Ophthalmadine.   Local treatment for herpetic eye infections: see **Idoxuridine.**
Opilon.   Peripheral vasodilator: see **Thymoxamine.**
Opticrom.   Eye drops for allergic conjunctivitis: see **Sodium cromoglycate.**
Optimax.   Antidepressant: see **Tryptophan.**
Opulets.   Single dose ophthalmic preparations.

Orabase. Topical inert protective application for skin and mucosae.

Orabolin. Anabolic steroid: see **Ethyloestrenol.**

Oradexon. Corticosteroid: see **Dexamethasone.**

Orahesive. Topical inert protective powder for skin and mucosae.

Oralcer. For oral ulcers: see **Clioquinol.**

Oraldene. Rinse for oral infections: see **Hexetidine.**

Orap. Tranquilliser: see **Pimozide.**

Orastrep (d). Antibiotic: see **Streptomycin.**

Oratrol. For glaucoma: see **Dichlorphenamide.**

Orbenin. Antibiotic: see **Cloxacillin.**

Organidin. Expectorant containing iodine.

Orgraine. Vasoconstrictor for migraine: see **Ergotamine, Caffeine, Hyoscyamine, Atropine sulphate, Phenacetin.**

Orisulf. Antibacterial: see **Sulphaphenazole.**

Orlest 21. Oral contraceptive: see **Ethinyloestradiol, Norethisterone.**

Orovite. Vitamin mixture: see **Thiamine, Riboflavine, Pyridoxine, Nicotinamide, Vitamin C.**

Ortho-Creme. Spermicidal cream: see **Ricinoleic acid, Sodium laurel sulphate, Nonoxynol**

Ortho-Forms. Spermicidal pessary.

Ortho-Gynol. Spermicidal jelly.

Ortho-Novin. Oral contraceptive: see **Mestranol, Norethisterone.**

Orthoxicol. Cough suppressant: see **Codeine, Methoxyphenamine.**

Orthoxine. Bronchodilator: see **Methoxyphenamine.**

Orudis. Non-steroid anti-inflammatory/analgesic: see **Ketoprofen.**

Ospolot. Anticonvulsant: see **Sulthiame.**

Ossopan. Source of calcium and fluoride for bone and dental states.

Otamidyl (d). Anti-infective ear drops: see **Propamidine, Diamidino-diphenylamine.**

Otopred. Antibacterial/anti-inflammatory ear drops: see **Chloramphenicol, Prednisolone, Thiomersal.**

Otoseptil. Antibiotic/anti-inflammatory ear drops: see **Neomycin, Tyrothricin, Hydrocortisone.**

Otosporin. Antibiotic/anti-inflammatory ear drops: see **Neomycin, polymyxin, Hydrocortisone.**

Ototrips. Antibiotic topical preparation for ear: see **Polymyxin, Bacitracin.**

Otrivine. Nasal decongestant: see **Xylometazoline.**

Ouabaine Arnaud. For treatment of cardiac failure and abnormal heart rhythms: see **Ouabaine.**

Ovanon. (d) Oral contraceptive: see **Mestranol, Lynoestrenol.**

Ovestin. Female sex hormone for deficiency states: see **Oestriol.**

Ovol. Antispasmodic: see **Dicyclomine.**

Ovran. Oral contraceptive: see **Ethinyloestradiol, Norgestrel.**

Ovranette. Oral contraceptive: see **Ethinyloestradiol, Norgestrel.**

Ovulen. Oral contraceptive: see **Mestranol, Ethynodiol.**

Ovysmen. Oral contraceptive: see **Ethinyloestradiol, Norethisterone.**

Oxydon. Antibiotic: see **Oxytetracycline.**

Oxymycin. Antibiotic: see **Oxytetracycline.**

P

Pabracort. Corticosteroid nasal insufflation: see **Hydrocortisone.**

Pacitron. Antidepressant: see **Tryptophan.**

Paedo-Sed. Hypnotic/analgesic: see **Dichloralphenazone, Paracetamol.**

Palaprin forte. Non-steroid, anti-inflammatory/analgesic: see **Aloxiprin.**

Palfium (c). Narcotic analgesic: see **Dextromoramide.**

Paludrine. Antimalarial: see **Proguanil.**

Pamergan (c). Premedication combination containing narcotic analgesic: see **Pethidine, Promethazine.**

Paminal. Sedative/antispasmodic: see **Hyoscine methobromide, Phenobarbitone.**

Pamine. Antispasmodic antacid: see **Hyoscine methobromide.**

Panadeine CO. Analgesic: see **Paracetamol, Codeine.**

Panadol. Analgesic: see **Paracetamol.**

Panar. For cystic fibrosis: see **Pancreatic enzymes.**

Panasorb. Analgesic: see **Paracetamol.**

Pancrex/Pancrex V/Pancrex V Forte. For use in pancreatic deficiency: see **Pancreatin.**

Panoxyl 5 and 10. Topical treatment for acne: see **Benzoyl peroxide.**

Parabal. Anticonvulsant/hypnotic: see **Phenobarbitone.**

Paracodol. Effervescent analgesic/antipyretic: see **Paracetamol, Codeine.**

Paradione. Anticonvulsant: see **Paramethadione.**

Paragesic. Effervescent analgesic/antipyretic/decongestant: see **Paracetamol, Pseudoephedrine, Caffeine.**

Parahypon. Analgesic/antipyretic: see **Paracetamol, Codeine, Caffeine, Phenolphthalein.**

Parake. Analgesic/antipyretic: see **Paracetamol, Codeine.**

Paralgin. Analgesic/antipyretic: see **Paracetamol, Codeine, Caffeine.**

Paramisan. Antituberculosis: see **P.A.S.**

Paramol-118. Analgesic: see **Paracetamol, Dihydrocodeine.**

Para-Seltzer. Effervescent analgesic/antipyretic: see **Paracetamol, Caffeine.**

Para-Thor-Mone. Hormone extract: see **Parathyroid hormone.**

Parentrovite. Anti-inflammatory/analgesic: see **Paracetamol, Phenylbutazone.**

Pardale. Analgesic/antipyretic: see **Paracetamol, Codeine, Caffeine.**

Parazolidin. High-potency parenteral vitamins for treatment of alcohol- or drug-induced psychosis and other debilitated conditions: see **Thiamine, Riboflavine, Pyridoxine, Nicotinamide, Vitamin C.**

Parfenac. Non-steroid anti-inflammatory cream: see **Bufexamac.**

Parlodel. Dopamine agonist: see **Bromocriptine.**

Parnate. Antidepressant monoamine oxidase inhibitor: see **Tranylcypromine.**

Paroven. Vitamin derivative for symptomatic treatment of aching associated with varicose veins: see **Troxerutin.**

Parstelin. Antidepressant monoamine oxidase inhibitor/tranquilliser: see **Tranylcypromine Trifluoperazine.**

Pasinah Preps. Anti-tuberculosis: see **P.A.S., Isoniazid.**

Pavacol. Cough suppressant: see **Papaverine, pholcodeine.**

Pavulon. Muscle relaxant: see **Pancuronium.**

Paynocil. Analgesic: see **Acetylsalicylic acid.** Formulated with **Glycine** to reduce gastric-irritation.

Pectomed. Cough suppressant: see **Ipecacuanha, Squill, Tolu, Ammonium acetate.**

Peganone. Anti-convulsant: see **Ethotoin.**

Penbritin. Antibiotic: see **Ampicillin.**

Penbritin KS. Antimicrobial/antidiarrhoeal: see **Ampicillin, Sulphadimidine, Kaolin.**

Penidural preps. Antibiotic: see **Benzathine penicillin.**

Penotrane pessaries. Anti-infective for vaginal infections: see **Hydrargaphen.**

Penthrane. Anaesthetic gas: see **Methoxyflurane.**

Pentostam. Antimony derivative: see **Stibogluconate sodium.**

Pentothal. Intravenous barbiturate anaesthetic: see **Thiopentone sodium.**

Pentovis. Sex hormone for post-menopausal symptoms: see **Quinestradol.**

Pentoxylon. Anti-anginal: see **Rauwolfia, Penterythritol tetranitrate.**

Pentral 80 Tempules. Anti-anginal (sustained-release formulation): see **Pentaerythritol tetranitrate.**

Pentrexyl. Antibiotic: see **Ampicillin.**

Pentrium (d). Anti-anginal: see **Pentaerythritol tetranitrate, Chlordiazepoxide.**

Peptard. Anticholinergic for peptic ulcers and abdominal colic: see **Hyoscyamine.**

Peptavlon. Hormone injection for tests of gastric secretion: see **Pentagastrin.**

Peralvex. For inflammation of the oral mucosa: see **Salicylic acid.**

Percorten M Crystules. Adrenocorticosteroid depot injection for treatment of adrenal insufficiency: see **Deoxycortone.**

Periactin. Serotonin antagonist for stimulation of appetite: see **Cyproheptadine.**

Perideca. Steroid/serotonin antagonist for allergic disorders including eczema: see **Dexamethasone, Cyproheptadine.**

Peritrate. Anti-anginal: see **Pentaerythritol tetranitrate.**

Peritrate SA. As Peritrate but in sustained-release formulation.

Peritrate with phenobarbitone. As peritrate plus **Phenobarbitone** as sedative.

Pernivit. Vitamins for chilblains: see **Acetomenapthone, Nicotinic acid.**

Pernomol. Topical treatment for chilblains: see **Chlorbutol, Phenol, Tannic acid, Soap spirit, Camphor.**

Peroidin. For hyperthryroidism: see **Potassium perchlorate.**

Perolysen (d). Antihypertensive: see **Pempidine.**

Persantin. For ischaemic heart disease: see **Dipyridamole.**

Pertofran. Antidepressant: see **Desipramine.**

Penthilorfan (c). Narcotic analgesic plus narcotic antagonist. Recommended as an analgesic where there is an obvious risk of respiratory depression, e.g. to the foetus during labour: see **Pethidine, Levallorphan.**

Petrolagar No. 1. Purgative: see **Liquid paraffin.**

Petrolagar No. 2. As Petrolagar No. 1 plus **Phenolphthalein.**

Pexid. Anti-anginal: see **Perhexilene.**

Phanodorm. Hypnotic: see **Cyclobarbitone.**

Phasal. Sustained-release tranquilliser: see **Lithium.**

Phazyme. For flatulence and dyspepsia: see **Simethicone, Pancreatin.**

Phenergan. Antihistamine for topical and systemic use: see **Promethazine.**

Phenergan Compound. Cough linctus/decongestant: see **Promethazine, Ipecacuanha, Potassium guaiacolsulphonate.**

Phenocitrain. Topical antiseptic for varicose ulcers or burns: see **Phenol, Lignocaine.**

Phenomet. Hypnotic/sedative plus emetic to prevent absorption in over-dosage: see **Phenobarbitone, Emetine.**

Phensedyl. Cough linctus/decongestant: see **Promethazine, Codeine, Ephedrine.**

Phiso-Med. Anti-infective for cleansing skin/hair: see **Hexachlorophane.**

Pholcomed. Cough linctus: see **Pholcodine, Papaverine.**

Pholtex. Cough linctus/decongestant: see **Pholcodine, Phenyltoloxamine.**

Phosphate-Sandoz. Effervescent **Phosphate** supplement for hyperparathyroidism and other bone disease.

Phospholine iodide. Preparations for open-angle glaucoma: see **Ecothiopate.**

Phyllocontin. For asthma and cardiac failure: see **Aminophylline.**

Physeptone (c). Narcotic analgesic: see **Methadone.**

Phytex. Topical antifungal paint: see **Boric acid, Tannic acid, Salicylic acid.**

Phytocil. Topical antifungal cream and powder: see **Phenoxypropanol, Chlorophenoxyethanol, Salicylic acid.**

Phytodermine. Topical antifungal powder: see **Methyl hydroxybenzoate, Salicylic acid.**

Pib/Pib plus. Aerosol bronchodilator: see **Isoprenaline, Atropine methonitrate.**

Pimafucin. Topical antifungal for skin, vagina or lungs (by inhalation): see **Natamycin.**

Pipanol (d). Anticholinergic for Parkinsonism: see **Benzhexol.**

Piptal. Anticholinergic for peptic ulcers: see **Pipenzolate.**

Piptalin. As Piptal plus **Simethicone** for abdominal colic.

Piriton. Antihistamine: see **Chlorpheniramine.**

Pitocin. Produces uterine contraction: see **Oxytocin.**

Pitressin. Hormone: see **Antidiuretic hormone.**

P.K. Aid 1 (b). Dietary aid for phenylketonuria.

Plaquenil. Anti-inflammatory: see **Hydroxychloroquine.**

Plastules with folic acid. For anaemia: see **Ferrous sulphate, Folic acid.**

Plastules with liver extract. Iron and liver extract for anaemia: see **Ferrous sulphate, Liver extracts.**

Plesmet. For iron-deficiency anaemia: see **Ferrous sulphate.**

Plex-Hormone. Male sex hormone replacement: see **Methyltestosterone, Deoxycortone, Ethinyloestradiol, Tocopheryl.**

Pollinex. Pollen extracts for desensitisation in asthma, hay fever.

Polyalk. Antacid: see **Dimethicone, Aluminium hydroxide.**

Polybactrin. Antibiotic: see **Polymyxin B, Bacitracin, Neomycin,**

Polycrol forte. Antacid: see **Simethicone, Magnesium hydroxide, Aluminium hydroxide.**

Polyfax. Antibiotic: see **Polymyxin B, Bacitracin.**
Polytar (b). Topical treatment for psoriasis: see **Coal tar.**
Polyvite. Vitamin mixture: see **Vitamin A, Calciferol, Thiamine, Riboflavine, Pyridoxine, Vitamin C, Nicotinamide, Pantothenic acid.**
Ponderax. Anti-obesity: see **Fenfluramine.**
Ponoxylan. Topical anti-infective/anti-inflammatory: see **Polynoxylin.**
Ponstan. Anti-inflammatory/analgesic: see **Mefenamic acid.**
Portagen (b). Dietary aid for lactose intolerance.
Posalfilin. Topical treatment for warts: see **Salicylic acid, Podophyllum.**
Potaba. Non-steroid anti-inflammatory: see **Potassium-*para*-aminobenzoate.**
Potensan. For loss of libido: see **Yohimbine, Strychnine, Amylobarbitone.**
Potensan forte. For loss of libido: see **Yohimbine, Methyltestosterone, Strychnine, Pemoline.**
Practo-Clyss. Purgative enema: see **Sodium phosphate.**
Pragmatar. Topical treatment for seborrhoea: see **Coal tar, Sulphur, Salicylic acid.**
Pramidex. Oral hypoglycaemic: see **Tolbutamide.**
Praxilene. Peripheral vasodilator: see **Naftidrofuryl.**
Preceptin (d). Spermicidal preparation.
Precortisyl. Corticosteroid: see **Prednisolone.**
Predenema. Corticosteroid enema: see **Prednisolone.**
Prednesol. Corticosteroid: see **Prednisolone.**
Predsol. Corticosteroid: see **Prednisolone.**
Predsol-N. Corticosteroid/antibiotic ear drops: see **Prednisolone, Neomycin.**
Pregaday. For anaemia of pregnancy: see **Ferrous fumarate, Folic acid.**
Pregestimil (b). Dietary aid for glucose, lactose, protein intolerance.
Pregfol. For anaemia of pregnancy: see **Ferrous sulphate, Folic acid.**
Pregnavite forte. For anaemia of pregnancy: see **Ferrous sulphate, Folic acid, Vitamin A, Vitamin B, Calciferol, Vitamin C.**
Pregnyl. Hormone: see **Gonadotrophin.**
Prehensol. Protective skin cream: see **Zinc salicylate.**
Premarin. Natural oestrogens from pregnant mare's urine for deficiency states.
Prenomiser (d). Bronchodilator aerosol: see **Isoprenaline.**
Prenomiser Plus. Bronchodilator aerosol: see **Isoprenaline, Atropine methonitrate.**
Pressimmune. Anti-human lymphocyte globulin for immunosuppression.
Pressurised Brovon. Bronchodilator aerosol: see **Adrenaline, Atropine methonitrate.**
Priadel. Antidepressant: see **Lithium salts.**
Primalan. Antihistamine: see **Mequitazine.**
Primodos. Sex hormones for amenorrhoea: see **Norethisterone, Ethinyloestradiol.**
Primogyn. Female sex hormone for deficiency states: see **Oestradiol.**
Primolut depot (d). For recurrent abortion: see **Hydroxyprogesterone.**
Primolut N. To reduce or postpone menstruation: see **Norethisterone.**
Primoteston depot. Male sex hormone for deficiency states: see **Testosterone.**

Primperan. Anti-emetic. Promotes gastric emptying: see **Metoclopramide.**

Prinalgin. Non-steroid anti-inflammatory: see **Alcofenac.**

Prioderm. Topical treatment for lice: see **Malathion.**

Pripsen. For threadworms, roundworms: see **Piperazine.**

Priscol. Peripheral vasodilator: see **Tolazoline.**

Pro-Actidil. Antihistamine: see **Tripolidine.**

Pro-Banthine. Anticholinergic used for antispasmodic and antacid effects: see **Propantheline.**

Proctofoam HC. Topical treatment for ano-rectal conditions: see **Hydrocortisone, Pramoxine.**

Proctosedyl. Local treatment for haemorrhoids: see **Hydrocortisone, Cinchocaine, Framycetin.**

Prodexin. Antacid: see **Aluminium glycinate, Magnesium carbonate.**

Prodoxol. Urinary anti-infective: see **Oxolinic acid.**

Progynova. Female sex hormone for deficiency states: see **Oestradiol.**

Proladone (c). Narcotic analgesic: see **Oxycodone.**

Prominal. Anticonvulsant: see **Methylphenobarbitone.**

Prondol. Antidepressant: see **Iprindole.**

Pronestyl. Antidysrhythmic: see **Procainamide.**

Propaderm. Topical corticosteroid: see **Beclomethasone.**

Proper-Myl. Yeast extract injection.

Prosobee (b). Dietary aid for lactose intolerance.

Prosol (b). Dietary aid for coeliac disease.

Prosparol (b). Calorie source.

Prostigmin. Anticholinesterase: see **Neostigmine.**

Proteolysed liver Pabyrn. Predigested beef liver for anaemia.

Prothiaden. Antidepressant: see **Dothiepin.**

Provera. For menstrual disorders and prevention of threatened abortion: see **Medroxyprogesterone.**

Provera 100 mg. High dose **Medroxyprogesterone** for treatment of endometrial carcinoma or hypernephroma.

Pro-Viron. Male sex hormone for deficiency states: see **Mesterolone.**

P.R. Spray. Rubefacient: see **Chlorofluoromethane.**

Pruvagol (d). Anti-infective pessaries for vaginal infections: see **Borax.**

Psoriderm preps. Topical treatments for psoriasis: see **Coal tar, Lecithin.**

Psoriderm-S. As Psoriderm plus **Salicylic acid.**

Psorox. Topical treatment for psoriasis: see **Coal tar.**

Pularin. Anticoagulant: see **Heparin.**

Pulmadil. Bronchodilator aerosol: see **Rimiterol.**

Pulmodrine. Cough linctus: see **Guaiphenesin, Methylephedrine.**

Puri-Nethol. Cytotoxic: see **Mercaptopurine.**

Puroverine inj. Hypotensive for pre-eclampsia: see **Veratrum.**

Pylura. Topical treatment for haemorrhoids: see **Adrenaline, Benzamine, Phenol.**

Pyopen. Antibiotic: see **Carbenicillin.**

Pyorex. Antiseptic toothpaste for mouth and gum infections: see **Acetarsol, Aminacrine.**

Pyridium. Analgesic for urinary tract: see **Phenazopyridine.**

Q

Quellada. Topical treatment for scabies and lice: see **Gamma-benzene hexachloride.**

Questran. For hypercholesterolaemia: see **Cholestyramine.**

Quinicardine. Antidysrhythmic: see **Quinidine.**

Quinoderm. Topical treatment for acne: see **Hydroxyquinoline, Benzoyl peroxide.**

Quinoderm with 1% hydrocortisone. As Quinoderm plus **Hydrocortisone.**

Quinoped. Topical antifungal for feet: see **Hydroxyquinoline, Benzoyl peroxide.**

Quixalin. Anti-amoebic for amoebic dysentery: see **Halquinol.**

Quotane. Topical anaesthetic ointment: see **Dimethisoquin.**

R

Rabro. Antacid: see **Bismuth subnitrate, Magnesium oxide, Calcium carbonate, Frangula, Liquorice.**

Rapidal. Hypnotic: see **Cyclobarbitone.**

Rastinon. Hypoglycaemic: see **Tolbutamide.**

Raudixin. Antihypertensive: see **Rauwolfia.**

Rautrax. Antihypertensive. As Raudixin plus **Hydroflumethiazide, Potassium chloride.**

Rautrax Sine K. Antihypertensive. As Rautrax but no **Potassium chloride.**

Rauwiloid. Antihypertensive: see **Rauwolfia.**

Rauwiloid + Veriloid. Antihypertensive: see **Rauwolfia, Veratrum.**

Razoxin. Antitumour: see **Razoxane.**

R.B.C. Topical cream for sunburn: see **Benzocaine, Calamine, Cholesterol, Phenylmercuric nitrate.**

Reactivan. C.N.S. stimulant and vitamins for tonic: see **Fencamfamin, Thiamine, Pyridoxine, Cyanocobalamin, Vitamin C.**

Reasec. Antidiarrhoeal: see **Diphenoxylate, Atropine sulphate.**

Redeptin. Depot injection tranquillisers/anti-psychotic: see **Fluspirilene.**

Redoxon. Vitamin: See **Vitamin C.**

Relefact LH-RH. For diagnostic use in delayed sexual development and failure of pituitary gland function: see **Gonadorelin.**

Remiderm. Topical corticosteroid/anti-infective: see **Triamcinolone, Halquinol.**

Remnos. Hypnotic: see **Nitrazepam.**

Remotic. Topical adrenocorticosteroid/anti-infective for outer ear: see **Triamcinolone, Halquinol.**

Rendells. Spermicidal pessary/foam: see **Nonoxynol.**

Resochin. Antimalarial/anti-inflammatory: see **Chloroquine.**

Resonium-A. Ion-exchange resin: see **Sodium polystyrene sulphonate.**

Retcin. Antibiotic: see **Erythromycin.**

Retin-A. Topical treatment for acne: see **Tretinoin.**

Revonal (c). Hypnotic: see **Methaqualone.**

Rheomacrodex. Plasma expander: see **Dextrans.**

Rheumajecta. Intramuscular enzyme injection for rheumatoid disease.

Rheumox. Non-steroid anti-inflammatory: see **Azapropazone.**

Rhinamid. Nasal decongestant/anti-inflammatory: see **Ephedrine, Sulphanilamide, Butacaine.**

Riddospas. Bronchodilator: See **Theobromine, Theophylline.**

Riddovydrin. Bronchodilator for inhalation: see **Adrenaline, Chlorbutol, Sodium nitrate, Papaverine, Atropine methonitrate, Vitamin C, Pituitary gland extract.**

Rifadin. Anti-tuberculosis: see **Rifampicin.**

Rifinah. Anti-tuberculosis: see **Rifampicin, Isoniazid.**

Rikospray antibiotic. Topical antibiotic spray: see **Neomycin,Bacitrac in, Polymixin B.**

Rikospray balsam. Aerosol spray for sore skin, e.g. nappy rash: see **Benzoin, Storax.**

Rikospray silicone. Aerosol spray for sore skin, e.g. bed sores: see **Cetylpyridinium, Dimethicone.**

Rimactane. Anti-tuberculosis: see **Rifampicin.**

Rimactazid. Anti-tuberculosis: see **Rifampicin, Isoniazid.**

Rinurel. Analgesic/decongestant for symptomatic relief of the common cold: see **Paracetamol, Phenylpropanolamine, Phenyltoloxamine.**

Ritalin (c). C.N.S. stimulant: see **Methylphenidate.**

Rite-Diet gluten free (b). Dietary substitute for **Gluten** sensitivity.

Rite-Diet protein free (b). Dietary substitute for protein intolerance, e.g. renal failure.

Rivotril. Anticonvulsant: see **Clonazepam.**

Ro-A-Vit. **Vitamin A** supplement.

Robaxin 750. Muscle relaxant: see **Methocarbamol.**

Robaxisal forte. As Robaxin 750 plus **Acetylsalicylic acid.**

Robinul. Anticholinergic for peptic ulcers: see **Glycopyrronium.**

Robitussin. Expectorant: see **Guaiphenesin.**

Robitussin A.C. Cough linctus: see **Guaiphenesin, Codeine, Pheniramine.**

Roccal. Skin antiseptic: see **Benzalkonium.**

Rogitine. For diagnostic test/treatment of phaeochromocytoma: see **Phentolamine.**

Rona-Slophyllin. Sustained release bronchodilator: see **Theophylline.**

Rondomycin. Antibiotic: see **Methacycline.**

Ronicol. Peripheral vasodilator: see **Nicotinyl tartrate.**

Ronyl. C.N.S. stimulant: see **Pemoline.**

Roter. Antacid: see **Magnesium carbonate, Bismuth subnitrate, Sodium bicarbonate, Frangula.**

Rotercholon (d). For biliary disorders: see **Fennel, Turmeric, Ox bile extract, Methyl salicylate.**

Rotersept. Antiseptic spray for prevention of mastitis during lactation: see **Chlorhexidine.**

Rovamycin. Antibiotic: see **Spiramycin.**
Rowachol. For biliary disorders. Mixture of essential oils: see **Menthol, Camphor, Eucalyptus.**
Rowatinex. For biliary disorders: Mixture of **Essential oils,** e.g. **Anethol, Camphor, Eucalyptus.**
Rubelix. Cough linctus: see **Pholcodine, Ephedrine.**
Ruthmol. Salt substitute for low-sodium diets: see **Potassium chloride.**
Rybarex. Bronchodilator aerosol. As Rybarvin plus **Methyl salicylate.**
Rybarvin. Bronchodilator aerosol: see **Atropine methonitrate, Adrenaline, Papaverine, Benzocaine, Posterior pituitary extract.**
Rynabond (d). Long acting decongestant: see **Phenylephrine, Pheniramine, Mepyramine.**
Rynacrom. Topical insufflation for allergic rhinitis: see **Sodium chromoglycate.**
Rythmodan. Antidysrhythmic: see **Disopyramide.**

S

Safapryn. Analgesic: see **Acetylsalicylic acid, Paracetamol.**
Safapryn-Co. As Safapryn plus **Codeine.**
Salactol. Topical treatment for warts: see **Salicylic acid, Lactic acid.**
Salaphene (d). Topic gel for acne: see **Salicylic acid, Resorcinol, Bithionol.**
Salazopyrin. For ulcerative colitis: see **Sulphasalazine.**
Salimed. Analgesic: see **Salicylamide.**
Salimed compound. Analgesic/sedative: see **Salicylamide, Mephenesin, Amylobarbitone.**
Salupres. Antihypertensive: see **Reserpine, Hydroclorothiazide, Potassium chloride.**
Saluric. Diuretic: see **Chlorothiazide.**
Salzone. Analgesic: see **Paracetamol.**
Sancos. Cough linctus: see **Pholcodine, Menthol.**
Sancos CO. Cough linctus/decongestant. As Sancos plus **Pseudoephedrine, Chlorpheniramine.**
Sandocal. Effervescent supplement for calcium deficiency states.
Sando-K. Effervescent potassium supplement. Mixture of **Potassium chloride** and **Potassium bicarbonate.** Provides potassium and chloride ions for absorption. Gastric irritation much less than with simple potassium chloride solution. Some irritation may still occur. Danger of hypercalcaemia if used in renal failure, treated by haemodialysis and ion exchange resins.
Sanomigran. Migraine prophylactic: see **Pizotifen.**
Saridone. Analgesic: see **Phenacetin, Propyphenazone, Caffeine.**
Saroten. Antidepressant: see **Amitriptyline.**
Saventrine. Slow-release sympathomimetic for heart block: see **Isoprenaline.**
Savlodil. Antiseptic for skin wounds or burns: see **Chlorhexidine, Cetrimide.**
Savlon hospital concentrate. Antiseptic: see **Chlorhexidine.**

Schericur. Topical corticosteroid/antiseptic: see **Hydrocortisone, Clemizole, Hexachlorophane.**

Scheriproct. Topical treatment for haemorrhoids: see **Prednisolone, Cinchocaine, Hexachlorophane, Clemizole.**

SDV. Desensitising vaccines for certain allergies, e.g. asthma.

Seboderm (b) (d). Shampoo for seborrhoeic dermatitis: see **Cetrimide, Cetyl alcohol.**

Secaderm salve. Topical treatment for chilblains, boils, bunions: see **Phenol, Terebene, Turpentine oil.**

Seconal sodium. Sedative/hypnotic: see **Quinalbarbitone.**

Sectral. Beta adrenoceptor blocker: see **Acebutalol.**

Sedatussin. Cough linctus: see **Cephaeline, Sodium benzoate, Squill, Menthol.**

Sedestran. Sex hormone/sedative for menopausal disorders: see **Stilboestrol, Phenobarbitone.**

Sedonan. Topical analgesic drops for painful ear infections: see **Phenazone, Chlorbutol.**

Selora. Salt substitute: see **Potassium chloride.**

Selsun (b). Shampoo for seborrhoeic dermatitis: see **Selenium sulphide.**

Selvigon (d). Cough linctus: see **Pipazethate.**

Senokot. Purgative: see **Senna.**

Seominal. Antihypertensive/sedative: see **Reserpine, Theobromine, Phenobarbitone.**

Septex No. 1. Soothing cream for sore skin, e.g. nappy rash: see **Boric acid, Zinc oleate, Zinc oxide.**

Septex No. 2. As Septex No. 1 plus anti-infective: see **Sulphathiazole.**

Septrin. Antibacterial: see **Cotrimoxazole.**

Serc. For Ménière's syndrome: see **Betahistine.**

Serenace. Tranquilliser: see **Haloperidol.**

Serenesil (d). Hypnotic: see **Ethchlorvynol.**

Serenid-D/Serenid-Forte. Anxiolytic: see **Oxazepam.**

Serpasil. Antihypertensive: see **Reserpine.**

Serpasil-Esidrex. Antihypertensive: see **Reserpine, Hydrochlorothiazide.**

Serpasil Esidrex K. Antihypertensive: see **Reserpine, Hydrochlorothiazide, Potassium chloride.**

SH 420. Sex hormone for inoperable carcinoma of the breast: see **Norethisterone.**

Sidros. Haematinic: see **Ferrous gluconate, Vitamin C.**

Silbe inhalant. Bronchodilator inhalation: see **Adrenaline, Atropine methonitrate, Papaverine, Hyoscine.**

Silberphylline. Bronchodilator, see **Diprophylline.**

Silderm. Topical corticosteroid/anti-infective: see **Triamcinolone, Neomycin.**

Silocalm (d). Antacid/anticholinergic for peptic ulceration: see **Aluminium hydroxide, Dimethicone, Propantheline.**

Siloxyl. Antacid: see **Aluminium hydroxide, Simethicone.**

Simeco. Antacid: see **Aluminium hydroxide, Simethicone.**

Simplene. Eye drops for glaucoma: see **Adrenaline.**

Sinaxar.   Muscle relaxant: see **Styramate.**

Sinemet and Sinemet-100.   Antiparkinsonian: see **Carbidopa, Laevodopa.**

Sinequan.   Anxiolytic/antidepressant: see **Doxepin.**

Sinetens.   Antihypertensive: see **Prazosin.**

Sinthrome.   Anticoagulant: see **Nicoumalone.**

Sintisone.   Corticosteroid: see **Prednisolone.**

Siopel.   Soothing, antiseptic cream for skin rashes: see **Dimethicone, Cetrimide.**

SK-65.   Analgesic: see **Dextropropoxyphene.**

Skefron.   Rubefacient aerosol: see **Dichlorodifluoromethane, trichlorofluoromethane.**

Skin testing solutions.   Allergen extracts used in skin testing for allergies.

Slow-Fe.   Slow-release haematinic: see **Ferrous sulphate.**

Slow-Fe folic.   Slow-release haematinic as Slow-Fe plus **Folic Acid.**

Slow-K.   Slow-release **Potassium chloride.**

Slow Sodium.   Slow-release **Sodium chloride.**

Slow-Trasicor.   Antihypertensive.   Sustained-release formulation of **Oxprenolol.**

Sno-Pilo.   Miotic eye drops for glaucoma: see **Pilocarpine.**

Sobee (b).   Lactose-free dietary supplement.

Sodium Amytal.   Hypnotic/sedative: see **Amylobarbitone.**

Sofradex.   Corticosteroid/anti-infective drops for use in eyes or ears: see **Dexamethasone, Framycetin, Gramicidin.**

Soframycin.   Topical anti-infective: see **Framycetin, Gramicidin.**

Soframycin inj. and tabs.   Antibiotic: see **Framycetin.**

Sofra-Tulle.   Gauze dressing with antimicrobial: see **Framycetin.**

Soliwax.   Softens ear wax: see **Dioctyl sodium sulphosuccinate.**

Solpadeine.   Soluble, effervescent analgesic: see **Paracetamol, Codeine, Caffeine.**

Solprin.   Soluble analgesic: see **Acetylsalicylic acid.**

Sol-Tercin.   Soluble analgesic/sedative: see **Acetylsalicylic acid, Butobarbitone.**

Solu-Cortef.   Corticosteroid injection: see **Hydrocortisone.**

Solu-Medrone.   Corticosteroid injection: see **Methylprednisolone.**

Sonalgin.   Analgesic/sedative: see **Phenacetin, Codeine, Butobarbitone.**

Sonergan.   Hypnotic/sedative: see **Promethazine, Butobarbitone.**

Soneryl.   Hypnotic/sedative: see **Butobarbitone.**

Sorbitol EGIC.   For intravenous feeding: see **Sorbitol.**

Sorbitrate.   Anti-anginal chewable tablets: see **Isosorbide dinitrate.**

Sotacor.   Beta adrenoceptor blocker: see **Sotalol.**

Spaneph Spansule (d).   Sustained release bronchodilator: see **Ephedrine.**

Sparine.   Tranquilliser anti-emetic: see **Promazine.**

Spasmonal.   Antispasmodic for gastro-intestinal or uterine spasm: see **Alverine.**

Spectraban (b).   Topical application for protection of skin from ultra-violet light.

Sporostacin.   Antiseptic cream for vaginal fungal infections: see **Benzalkonium.**

Sprilon.   Soothing, protective skin cream: see **Dimethicone, Zinc oxide.**
Stabillin V-K.   Antibiotic: see **Phenoxymethylpenicillin.**
Stannoxyl.   Oral treatment for boils, carbuncles, containing tin oxide.
Staycept.   Spermicidal contraceptive pessary: see **Nonoxynol.**
STD inj.   Schleroses varicose veins: see **Sodium tetradecyl sulphate.**
Steclin.   Antibiotic: see **Tetracycline.**
Stecsolin.   Antibiotic: see **Oxytetracycline.**
Stelabid.   For reducing gastric motility, secretion, anxiety: see **Trifluopera-zine, Isopropamide.**
Steladex (c).   Tranquilliser/stimulant: see **Trifluoperazine, Dexamphetamine.**
Stelazine.   Tranquilliser: see **Trifluoperazine.**
Stemetil.   Tranquilliser/anti-emetic/antivertigo: see **Prochlorperazine.**
Sterogyl-15.   For calcium disorders: see **Calciferol.**
Steroxin.   Topical anti-infective: see **Chlorquinaldol.**
Ster-Zac (b).   Topical anti-infective: see **Hexachlorophane.**
Stie-Lasan.   Topical treatments for psoriasis: see **Dithranol, Salicylic acid.**
Stomogel (b).   Stomal deodorant gel: see **Chlorhexidine, Benzalkonium.**
Strepsils.   Lozenges for minor oral infections: see **Dichlorobenzyl alcohol, Amylmetacresol.**
Streptotriad.   Antibiotic/antibacterial: see **Streptomycin, Sulphadiazine, Sulphathiazole, Sulphadimidine.**
Stromba.   Anabolic steroid: see **Stanozolol.**
Stugeron.   Anti-emetic antivertigo: see **Cinnarizine.**
Sublimaze (c).   Narcotic analgesic: see **Fentanyl.**
Sucrets.   Lozenges for minor oral infections: see **Hexylresorcinol.**
Sudafed.   Decongestant: see **Pseudoephedrine.**
Suleo (b).   Shampoo for head lice: see **Carbaryl.**
Sulfamylon.   Anti-infective cream for burns: see **Mafenide.**
Sulfapred.   Antibacterial/corticosteroid eyes drops: see **Sulphacetamide, Prednisolone.**
Sulfasuxidine.   Antibacterial: see **Succinylsulphathiazole.**
Sulfex.   Nasal antibacterial/decongestant: see **Sulphathiazole, Hydroxy-amphetamine.**
Sulfomyl.   Antibacterial eye drops: see **Mafenide.**
Sulphamagna.   Antibacterial/antidiarrhoeal: see **Attapulgite, Streptomycin, Sulphadiazine, Phthalylsulphathiazole.**
Sulphamezathine.   Antibacterial: see **Sulphadimidine.**
Sulphatriad.   Antibacterial combination: see **Sulphathiazole, Sulphadiazine, Sulphamerazine.**
Sultrin vaginal preps.   Local antibacterial combination: see **Sulphathiazole, Sulphacetamide, Sulphanilamide.**
Surbex T.   For vitamin B and C deficiency: see **Thiamine, Riboflavine, Nicotinamide, Pyridoxine, Vitamin C.**
Surmontil.   Antidepressive: see **Trimipramine.**
Suscardia.   Sympathomimetic amine: see **Isoprenaline.**
Sustac.   Anti-anginal: see **Glyceryl trinitrate.**
Sustamycin.   Antibiotic: see **Tetracycline.**

Sustanon. Male sex hormone for deficiency states or inoperable breast carcinoma: see **Testosterone.**
S.V.C. Pessary for vaginal infections: see **Acetarsol.**
Syl. Protective skin cream: see **Dimethicone, Benzalkonium.**
Sylopal. Antacid: see **Dimethicone, Magnesium oxide, Aluminium hydroxide.**
Symmetrel. Antiparkinsonian: see **Amantadine.**
Sympatol. Vasoconstrictor: see **Oxedrine.**
Synacthen. Synthetic corticotrophic injection: see **Tetracosactrin.**
Synadrin. Anti-anginal: see **Prenylamine.**
Synalar. Corticosteroid: see **Fluocinolone.**
Synandone. Corticosteroid: see **Fluocinolone.**
Synergel. Antacid: see **Aluminium antacids.**
Synkavit. For prothrombin deficiency: see **Menadiol.**
Synogist. Shampoo for dandruff.
Synthamin. Amino acid and electrolyte sources for intravenous feeding.
Syntocinon. Synthetic pituitary hormone: see **Oxytocin.**
Syntometrine. Contracts uterine muscle: see **Ergometrine, Oxytocin.**
Syntopressin. Synthetic pituitary hormone: see **Lypressin.**
Syrtussar. Cough suppressant: see **Dextromethorphan, Pheniramine.**
Sytron. For iron-deficiency anaemia: see **Sodium iron edetate.**

T

Tace. Sex hormone for menopausal symptoms, suppression of lactation, prostatic carcinoma: see **Chlorotrianisene.**
Tachostyptan. Promotes clotting: see **Thromboplastin.**
Tacitin. Tranquilliser: see **Benzoctamine.**
Tagamet. Gastric histamine receptor blocker; reduces acid secretion: see **Cimetidine.**
Talpen. Antibiotic: see **Talampicillin.**
Tampovagan. Pessaries containing **Stilboestrol** or **Neomycin** for vaginal complaints.
Tandacote. Non-steroid anti-inflammatory, enteric-coated to reduce gastric irritation: see **Oxyphenbutazone.**
Tandalgesic. Non-steroid anti-inflammatory/analgesic: see **Oxyphenbutazone, Paracetamol.**
Tanderil. Anti-inflammatory: see **Oxyphenbutazone.**
Tanderil Alka. Anti-inflammatory/antacid: see **Oxyphenbutazone, Aluminium hydroxide, Magnesium trisilicate.**
Taractan. Tranquilliser: see **Chlorprothixene.**
Tarband. Zinc and **Coal tar** bandage for eczema.
Tarcortin. Corticosteroid cream: see **Hydrocortisone, Coal tar.**
Taumasthman. Bronchodilator: see **Theophylline, Phenazone, Caffeine, Ephedrine, Atropine sulphate.**
Tavegil. Antihistamine: see **Clemastine.**
Taxol. Purgative: see **Pancreatic enzymes, Aloes.**

Tedral. Bronchodilator: see **Diprophylline, Ephedrine, Phenobarbitone.**
Teevex. Topical antipruritic: see **Crotamiton, Halopyramine.**
Tegretol. Anticonvulsant: see **Carbamazepine.**
Temetex. Corticosteroid cream: see **Diflucortolone.**
Tenavoid. Diuretic/tranquilliser for pre-menstrual syndrome: see **Bendrofluazide, Meprobamate.**
Tenormin. Beta adrenoceptor blocker: see **Atenolol.**
Tensilon. Diagnostic for myasthenia gravis: see **Edrophonium.**
Tenuate. Anti-obesity: see **Diethylpropion.**
Tercin. Analgesic/sedative: see **Acetylsalicylic acid, Butobarbitone.**
Tercoda. Cough suppressant: see **Codeine.**
Teronac. Anti-obesity: see **Mazindol.**
Terpalin. Cough suppressant: see **Codeine.**
Terpoin. Cough suppressant: see **Codeine.**
Terra-Bron. Antibiotic/bronchodilator: see **Oxytetracycline, Ephedrine.**
Terra-Cortril. Antibiotic/corticosteroid: see **Oxytetracycline, Hydrocortisone.**
Terramycin. Antibiotic: see **Oxytetracycline.**
Tertroxin. Thyroid hormone: see **Liothyronine.**
Testoral. Male sex hormone: see **Testosterone.**
Tetmosol (b). Topical treatment for scabies: see **Monosulfiram.**
Tetrabid. Antibiotic: see **Tetracycline.**
Tetrachel. Antibiotic: see **Tetracycline.**
Tetracyn. Antibiotic: see **Tetracycline.**
Tetralysal. Antibiotic: see **Lymecycline.**
Tetrazets. Antibiotic lozenges for throat infection: see **Bacitracin, Tyrothricin, Neomycin, Benzocaine.**
Tetrex. Antibiotic: see **Tetracycline.**
Tetrex PMT. Antibiotic: see **Rolitetracycline.**
T.H.A. For reversal of non-depolarising muscle relaxants: see **Tacrine.**
Thalamonal (c). Narcotic analgesic/tranquilliser combination: see **Fentanyl, Droperidol.**
Thalazole. Antibacterial: see **Phthalylsulphathiazole.**
Thean. Bronchodilator: see **Proxyphylline.**
Theodrox. Bronchodilator/antacid: see **Aminophylline, Aluminium hydroxide.**
Theogardenal. Sedative/vasodilator: see **Phenobarbitone, Theobromine.**
Theograd. Bronchodilator: see **Theophylline.**
Theominal. Sedative/vasodilator: see **Phenobarbitone, Theobromine.**
Theo-Nar. Bronchodilator: see **Theophylline, Noscapine.**
Theophorin (d). Antihistamine: see **Phenindamine.**
Thiaver. Antihypertensive combination: see **Veratrum, Thiazide.**
Thiazamide. Antibacterial: see **Sulphathiazole.**
Thoracin. Rubifacient: see **Nicotinic acid, Salicylic acid.**
Thovaline. Skin protective: see **Zinc oxide.**
Throsil. Lozenges for oral infections: see **Benzalkonium, Amethocaine.**
Thylin. Non-steroid anti-inflammatory/analgesic: see **Nifenazone.**
Thyropit. Combination of crude thyroid and anterior pituitary hormones.

Thytropar.   In diagnosis of thyroid disorder: see **Thyrotrophin.**
Tiglyssin.   Antispasticity: see **Tigloidine.**
Timodine.   Antifungal/corticosteroid cream: see **Nystatin, Hydrocortisone, Benzalkonium, Dimethicone.**
Tinaderm.   Topical anti-fungal: see **Tolnaftate.**
Tineafax.   Topical anti-fungal: see **Zinc undecenoate, Zinc naphthenate.**
Titralac.   Antacid: see **Calcium carbonate.**
Tixylix.   Cough suppressant mixture: see **Promethazine, Pholcodine, Phenylpropanolamine.**
Tofranil.   Antidepressant: see **Imipramine.**
Tolanase.   Oral hypoglycaemic: see **Tolazamide.**
Tolnate.   Anti-emetic: see **Prothipendyl.**
Toniron.   For iron deficiency anaemia: see **Ferrous sulphate.**
Tonivitan A & D syrup.   'Tonic': see **Vitamin A, Calciferol, Ferric ammonium citrate.**
Tonivitan B syrup.   'Tonic': see **Thiamine, Riboflavine, Pyridoxine, Nicotinamide, strychnine.**
Tonivitan caps.   Vitamins for deficiency states: see **Vitamin A, Thiamine, Nicotinic acid, Vitamin C, Calciferol.**
Tonsillin.   Antibiotic for throat infection: see **Phenoxymethylpenicillin, Benzalkonium.**
Topilar.   Topical corticosteroid: see **Fluclorolone.**
Topisone.   Corticosteroid cream: see **Hydrocortisone.**
Topisone C.   Topical corticosteroid/antibacterial: see **Hydrocortisone, Clioquinol.**
Torecan.   Anti-emetic/antivertigo: see **Thiethylperazine.**
Tosmilen eye drops.   For glaucoma: see **Demecarium.**
Totolin.   Decongestant: see **Phenylpropanolamine.**
Totomycin.   Antibiotic: see **Tetracycline.**
Trancopal.   Tranquilliser: see **Chlormezanone.**
Trandate.   Antihypertensive: see **Labetalol.**
Transvasin.   Rubefacient: see **Salicylic acid, Nicotinic acid.**
Tranxene.   Tranquilliser: see **Potassium clorazepate.**
Trasicor.   Beta adrenoceptor blocker: see **Oxprenolol.**
Trasylol.   Used in acute pancreatitis: see **Aprotinin.**
Travasol.   Intravenous nutrient.
Tremonil.   Anti-tremor: see **Methixene.**
Trentadil.   Bronchodilator: see **Bamifylline.**
Trental.   Peripheral vasodilator: see **Oxpentifylline.**
Treosulfan.   Cytotoxic: see **Threitol dimethane sulfonate.**
Trescatyl.   Anti-tuberculosis: see **Ethionamide.**
Trescazide.   Anti-tuberculosis: see **Ethionamide.**
Trevintix.   Anti-tuberculosis: see **Prothionamide.**
Tri-Adcortyl.   Topical corticosteroid/antibiotic: see **Triamcinolone, Nystatin, Neomycin, Gramicidin.**
Tricaderm.   Topical corticosteroid: see **Triamcinolone, Salicylic acid, Benzalkonium.**
Tricloryl.   Hypnotic: see **Triclofos.**

165

Tridesilon. Topical corticosteroid: see **Desonide.**
Tridione. Anticonvulsant: see **Troxidone.**
Trinuride. Anticonvulsant: see **Pheneturide, Phenytoin, Phenobarbitone.**
Triocos. Cough suppressant/decongestant: see **Pholcodine, Pseudoephedrine, Chlorpheniramine.**
Triogesic. Decongestant/analgesic: see **Paracetamol, Phenylpropanolamine.**
Triominic. Decongestant/antihistamine: see **Phenylpropanolamine, Mepyramine, Pheniramine.**
Triostam. Used in schistosomiasis: see **Sodium antimonylgluconate.**
Triotussic. Decongestant/antihistamine/analgesic: see **Phenylpropanolamine, Mepyramine, Pheniramine, Noscapine, Paracetamol.**
Triperidol. Tranquilliser: see **Trifluperidol.**
Triplopen. Sustained-action antibiotic: see **Benethamine penicillin, Procaine penicillin, Benzylpenicillin.**
Triptafen DA/Forte/Minor. Antidepressant/tranquilliser: see **Amitriptyline, Perphenazine.**
Triscal (d). Antacid: see **Calcium carbonate, Magnesium carbonate.**
Trivax. Triple vaccination against diphtheria, tetanus and pertussis.
Trobicin. Long-acting, single-dose antibiotic for gonorrhoea: see **Spectinomycin.**
Tromexan. Anticoagulant: see **Ethyl biscoumacetate.**
Trophysan. Essential and non-essential amino acids, minerals and vitamins for intravenous nutrition.
Tropium. Anxiolytic: see **Chlordiazepoxide.**
Tryptizol. Tricyclic antidepressant: see **Amitriptyline.**
Trypure Novo. Enzyme for topical use in wounds and cavities to hasten removal of blood clots and slough.
Tubarine. Muscle relaxant: see **Tubocurarine.**
Tuberculin Tine Test. Intradermal injection test for tuberculosis: see **Tuberculin.**
Tuinal. Hypnotic: see **Amylobarbitone, Quinalbarbitone.**
Tussifans. Cough linctus: see **Belladonna extract, Potassium citrate, Ipecacuanha, Squill.**
Tyrimide. Anticholinergic for gastro-intestinal colic and peptic ulcers: see **Isopropamide.**
Tyrosolven. Antiseptic/local anaesthetic throat lozenges: see **Cetylpyridinium, Tyrothricin, Benzocaine.**
Tyrozets. Local anaesthetic throat lozenges: see **Tyrothricin, Benzocaine.**

U

Ubretid. Anticholinesterase: see **Distigmine.**
Ulcedal. For peptic ulcers: see **Deglycyrrizinsed liquorice.**
Ultracortenol. Corticosteroid for intra-articular injection in arthritis: see **Prednisolone.**

Ultradil.   Topical corticosteroid for eczema: see **Fluocortolone.**

Ultralanum.   Topical corticosteroid/anti-infective: see **Fluocortolone, Clemizole, Hexachlorophane.**

Ultrandren.   Male sex hormone for deficiency states or inoperable breast carcinoma: see **Fluoxymesterone.**

Ultrapen.   Antibiotic: see **Propicillin.**

Ultraproct.   Local treatment for haemorrhoids: see **Fluocortolone, Cinchocaine, Clemizole, Hexachlorophane.**

Unguentum.   Protective cream for use on skin.   May be used as vehicle for drugs.

Unidiarea.   Antibacterial/antidiarrhoeal: see **Neomycin, Clioquinol, Attapulgite.**

Uniflu plus Gregovite C.   For symptomatic treatment of common cold: see **Diphenhydramine, Paracetamol, Caffeine, Phenylephrine, Codeine, Vitamin C.**

Unimycin.   Antibiotic: see **Oxytetracycline.**

Uniroid.   Local treatment for haemorrhoids: see **Hydrocortisone cinchocaine, Neomycin, Polymyxin B.**

Uraband.   Impregnated bandage for local treatment of exudative eczema: see **Zinc oxide, Urethane, Ichthammol.**

Uracil Mustard.   Cytotoxic: see **Uramustine.**

Urantoin.   Urinary antiseptic: see **Nitrofurantoin.**

Ureaphil.   Diuretic for intravenous infusion: see **Urea.**

Urispas.   Anticholinergic antispasmodic for urinary tract colic: see **Flavoxate.**

Urolucosil.   Urinary anti-infective: see **Sulphamethizole.**

Uromide.   Antibacterial/analgesic for painful urinary tract infections: see **Sulphacarbamide, Phenazopyridine.**

Uropol.   Urinary anti-infective: see **Tetracycline, Sulphamethizole, Phenazopyridine.**

Uteplex.   Muscle relaxant: see **Uridine triphosphoric acid.**

Uticillin.   Urinary anti-infective: see **Carfecillin.**

Uvistat (b).   Topical application for protection of skin from ultraviolet light: see **Mexenone.**

V

Valium.   Anxiolytic: see **Diazepam.**

Valledrine.   Cough linctus: see **Trimeprazine, Pholcodine, Ephedrine.**

Vallergan.   Antihistamine: see **Trimeprazine.**

Vallestril.   Sex hormone for menopausal disorders or suppression of prostatic carcinoma: see **Trimeprazine, Menthol, Phenylpropanolamine, Guaiphenesin, Sodium citrate, Ipecacuanha.**

Vallex.   Cough Linctus: see **Trimeprazine, Phenylpropanolamine, Guaiphenesin, Ipecacuanha.**

Valoid.   Antihistamine: see **Cyclizine.**
Vamin.   Amino acids and carbohydrate for intravenous nutrition.
Vanair.   Topical treatment for acne: see **Benzoyl peroxide, Sulphur.**
Vancocin.   Antibiotic: see **Vancomycin.**
Vanquin.   For threadworms: see **Viprynium.**
Varidase.   Enzymes for topical use in the removal of fibrinous or blood clots: see **Streptokinase, Streptodornase.**
Variotin.   Topical antifungal: see **Pecilocin.**
Vascardin.   Anti-anginal: see **Isosorbide dinitrate.**
Vasculit.   Peripheral vasodilator: see **Bamethan.**
Vascutonex.   Rubefacient: see **Diethylamine salicylate, Glycol salicylate.**
Vasocon-A.   Eye drops for minor irritations: see **Antazoline, Naphazoline, Boric acid.**
Vasocort (d).   Nasal spray for allergic rhinitis: see **Hydrocortisone, Hydroxy-amphetamine, Phenylephrine.**
Vasogen.   Soothing, protective cream for sore skin: see **Dimethicone, Zinc oxide, Calamine.**
Vasolastine.   Enzymes of lipid metabolism for vascular disorders and disorders of fat metabolism.
Vasopred.   Anti-inflammatory/anti-allergic eye drops: see **Prednisolone, Phenylephrine.**
Vasosulf.   Antibacterial eye drops: see **Sulphacetamide.**
Vasotran.   Peripheral vasodilator: see **Isoxsuprine.**
Vastarel.   Anti-anginal: see **Trimetrazine.**
Vasylox (d).   Local, nasal decongestant: see **Methoxamine.**
Vatensol.   Antihypertensive: see **Guanoclor.**
V-Cil-K.   Antibiotic: see **Phenoxymethylpenicillin.**
V-Cil-K Sulpha.   Antibacterial: see **Phenoxymethylpenicillin, Sulphadimidine.**
Veganin.   Analgesic: see **Acetylsalicylic acid, Paracetamol, Codeine.**
Velactin (b).   Low-sucrose food substitute for milk intolerance.
Velbe.   Cytotoxic: see **Vinblastine.**
Veldopa.   Antiparkinsonian: see **Laevodopa.**
Velosef.   Antibiotic: see **Cephradine.**
Ventolin.   Sympathomimetic bronchodilator: see **Salbutamol.**
Veracolate.   Purgative: See **Cascara, Phenolphthalein, Bile salts, Capsicum.**
Veractil.   Tranquillisers: see **Methotrimeprazine.**
Veracur.   Topical treatment for warts: see **Formaldehyde.**
Verdiviton elixir.   Vitamin mixture: see **Cyanocobalamin, Pantothenic acid, Nicotinamide, Riboflavine, Pyridoxine, Thiamine.**
Veriloid.   Antihypertensive: see **Rauwolfia.**
Veriloid VP.   Antihypertensive/sedative. As Veriloid plus **Phenobarbitone.**
Verkade (b).   **Gluten**-free biscuits for gluten-sensitive bowel disorders.
Vermox.   For threadworms, whipworms, roundworms, hookworms: see **Mebendazole.**
Vertigon Spansule.   Sustained-release anti-emetic: see **Prochlorperazine.**
Vibramycin.   Antibiotic: see **Doxycycline.**
Vibriomune.   Cholera vaccine.

Vibrocil. Nasal decongestant/antibiotic: see **Dimethindene, Phenylephrine, Neomycin.**

Vi-Daylin. Vitamin mixture: see **Vitamin A, Vitamin D, Thiamine, Riboflavine, Vitamin C, Nicotinamide, Pyridoxine.**

Vidopen. Antibiotic: see **Ampicillin.**

Villescon. 'Tonic': see **Prolintane, Thiamine, Riboflavine, Pyridoxine, Nicotinamide, Vitamin C.**

Vioform. Topical anti-infective: see **Clioquinol.**

Virvina. 'Tonic': see **Thiamine, Riboflavine, Pyridoxine, Nicotinamide.**

Visclair. Reduces mucous viscosity: see **Methylcysteine, Carbicysteine.**

Vi-Siblin. Purgative: see **Ispaghula.**

Visken. Beta adrenoceptor blocker: see **Pindolol.**

Vita-E. Vitamin E: see **Tocopheryl.**

Vitavel syrup. Vitamin mixture: see **Vitamin A, Calciferol, Thiamine, Vitamin C.**

Vivalan. Antidepressant: see **Viloxazine.**

Vivonex (b). Dietary supplement of amino acids, glucose, fat, vitamins and minerals.

Volital. C.N.S. stimulant: see **Pemoline.**

Vortel. Bronchodilator: see **Clorprenaline, Ethomoxane, Methapyrilene.**

W

Waxsol. Drops to soften ear wax: see **Dioctyl sodium sulphosuccinate.**

'WB' Warfarin. Oral anticoagulant: see **Warfarin.**

Welldorm. Hypnotic: see **Dichloralphenazone.**

Whatman sodium cellulose phosphate. Reduces calcium absorption: see **Sodium cellulose phosphate.**

Wright's vaporizer. Inhalation for nasal, bronchial congestion: see **Chlorocresol.**

X

Xerymenex. Softens ear wax: see **Triethanolamine.**

X-Prep. Preradiography purgative: see **Senna.**

Xylocaine. Local anaesthetic: see **Lignocaine.**

Xylocard. Local anaesthetic: see **Lignocaine.**

Xylodase. Topical local anaesthetic: see **Lignocaine.**

Xyloproct. Local anaesthetic/corticosteroid for anal conditions: see **Lignocaine, Hydrocortisone.**

Xylotox preps. Local anaesthetic: see **Lignocaine, Adrenaline.**

Y

Yomesan. For tapeworms: see **Niclosamide.**
Yutopar. Uterine relaxant: see **Ritodrine.**

Z

Zactipar. Analgesic: see **Ethoheptazine, Paracetamol.**
Zactirin. Analgesic: see **Ethoheptazine, Acetylsalicylic acid.**
Zarontin. Anti-epileptic: see **Ethosuximide.**
Zaroxolyn. Diuretic: see **Metolazone.**
Zeasorb. Powder for excessive perspiration: see **Chloroxylenol.**
Zinamide. Anti-tuberculosis: see **Pyrazinamide.**
Zincaband. Zinc paste bandage.
Zincfrin. Eye drops for conjunctival irritation: see **Phenylephrine, Zinc sulphate.**
Zyloric. For gout: see **Allopurinol.**
Zypanar. **Pancreatic enzymes** for deficiency states.

Warwickshire School of Nursing
Nurse Education Centre
Central Hospital
Near Warwick, Warwickshire CV35 7EB